Successful RFPs in Construction

MANAGING THE REQUEST FOR PROPOSAL PROCESS

Successful RFPs in Construction

MANAGING THE REQUEST FOR PROPOSAL PROCESS

Richard T. Fria

McGraw-Hill

New York | Chicago | San Francisco | Lisbon | London | Madrid | Mexico City
Milan | New Delhi | San Juan | Seoul | Singapore | Sydney | Toronto

The McGraw·Hill Companies

Library of Congress Cataloging-in-Publication Data

Fria, Richard T.
    Successful RFPs in construction : managing the request for proposal process / Richard
T. Fria.
      p.   cm.
    ISBN 0-07-144909-4
    1. Buildings—Specifications. 2. Construction industry—Subcontracting. 3. Requests
for proposals (Public contracts) 4. Contractors—Selection and appointment. I. Title.

TH425.F65   2005
692'.8—dc22

2005041498

1 2 3 4 5 6 7 8 9 0 DOC/DOC 0 1 0 9 8 7 6 5

ISBN 0-07-144909-4

*The sponsoring editor for this book was Cary Sullivan, the editing supervisor was Stephen M.
Smith, and the production supervisor was Pamela A. Pelton. It was set in ITC Cheltenham by
Cindy LaBreacht. The art director for the cover was Anthony Landi.*

Printed and bound by RR Donnelley.

McGraw-Hill books are available at special quantity discounts to use as premiums and sales
promotions, or for use in corporate training programs. For more information, please write
to the Director of Special Sales, McGraw-Hill Professional, Two Penn Plaza, New York, NY
10121-2298. Or contact your local bookstore.

 This book is printed on recycled, acid-free paper containing a minimum of 50%
recycled, de-inked fiber.

This book is dedicated to Heather O'Neill:

angel, pure heart, and shining soul.

Through her grace, light, and love,

this book came to fruition, and for this

I will be eternally grateful.

Contents

The View from Here

The construction industry is continually evolving. Change is the norm, and it permeates all facets of the construction process—from feasibility, to design, to construction, and finally, to occupancy. As the project progresses, it never stops changing, growing, and maturing, and ultimately it becomes reality in the form of the sum of all the process parts.

Yet the development continuum of a project is only one aspect of change. After 35 years in this business, I continue to be grateful for a career in an evolving industry. Development professionals—including architect, engineer, constructor, developer, owner, user, and a cast of hundreds—are fortunate to participate in projects that have a beginning, an end, and a lasting finished product of which to be proud.

With every project comes a new team of participants, new design challenges, new materials, new construction techniques, and new technology, all of which represent a fresh approach, an expanded professional network, and another beginning and end. Threading the same nut on the same bolt on an assembly line day after day would never work for me.

I have been fortunate to have teamed up with so many talented professionals, producing permanent recognizable landmarks on the skyline. I believe a can-do approach, an open-minded philosophy, and working with enthusiasm are integral to ensuring a positive outcome. I am continually astonished at how this personal approach to both work and life attracts like-minded teammates. And when the project is complete, the champagne corks are popping, and the happy owner is beaming, every team member from laborer to design professional can point out the product to family and friends and say, "That's my building!"

I was born to build. I've known this passion all my life. The forts I built out of scrap from the neighborhood construction project represented challenge and creativity and fulfilled my need to hammer nails, saw wood, smell sawdust, and finally occupy the fruit of my labor. And of course I always had help, friends who shared the experience, the sense of accomplishment, and the pride. Little did I know then just how special is the team approach.

I majored in business in college, but my calling for construction moved me to take a job as a brick carrier for a father-son home-building bricklayer team. Every afternoon break, the son would drive to the store and return with a six-pack of beer. We would sit and indulge in two beers each, connect with one another, and then complete the day's work. Now that was one heck of a college summer job! And I never forgot the connection from that team camaraderie.

My journey has included work as a laborer, carpenter, foreman, superintendent, project manager, construction executive, and, yes, consultant. I worked in the boom years of the early 1970s in the Colorado ski resorts of Aspen, Vail, Steamboat Springs, Breckenridge, and Crested Butte. I built a log lodge on a 7000-acre ranch in the remote Bitterroot Valley of Montana. I managed the construction of the computer lab that developed the software for the Space Shuttle and the construction of production and test facilities for top-secret defense satellites. I've managed the construction of high-rises, electronic manufacturing facilities, biotechnology R&D labs, historic landmark renovations, hotels, and destination retail and entertainment centers. I've been blessed with a diverse, opportunity-filled career and, as a result, have learned just how much I don't know. But the one thing that has stuck with me throughout is the value of teamwork.

The construction industry's very nature is teamwork. So many hands touch a project from inception to completion that they simply cannot be counted. Friends ask me, "How does a high-rise get built?" I contemplate this, become completely bewildered, look at them, and say, "I really don't know. Somewhere near the end a miracle occurs." Yes, at some point, the project is magically transformed from individual pieces of steel, wood, stone, and glass into "a building"—the sum of all its parts.

Of course I know the reality is that the project gets built through the orchestrated efforts of many participants and the application of intense thought, planning, and execution. What amazes me is that all these parts that make up the whole come together in this one place from all over the world, touched by multitudes. The aluminum window mullions may be fabricated locally, but the paint may be produced in France, the aluminum ore mined in Mexico, and the extrusions manufactured in Canada. The wood-paneling veneer may be from exotic Madagascar, from trees felled by Madagascarites, while the plywood backing may be produced in the southeast United States and the adhesives in China. Every component of the building has a story, of families from foreign lands, of boat captains and truck drivers, of millwrights, accountants, and factory owners.

I realized one day that "the miracle" occurs when the myriad products show up at the job site, manufactured and crafted to precise tolerances, finished in the intended colors and shapes, and at precisely the time required to accommodate a complex schedule. It is an orchestra that stretches around the world, encompassing instruments and players making one incredibly beautiful sound! It is the sound of construction.

What I take away from this is the criticality of teamwork—a necessary component for every successful project, from the smallest tenant improvement to the Golden Gate Bridge. "It's the team" is what I tell myself when I contemplate the miracle. And that is why this book presents a collaborative team approach to selecting a contractor. In the end, when the miracle needs to happen, it's the team that will make the difference.

The process presented in this book is a proven method, developed over many years and through many projects. As I train others, I am realizing that the process, like the industry itself, will continue to evolve. I look forward to learning much, and to the continued adventure.

Acknowledgments

I am grateful for those who have taught me and offered opportunity, encouragement, feedback, and enthusiasm for the writing of this book:

➤ My mentors, Ray Hews and Bill Lewis, who led by example and provided lifetime opportunities for personal and career growth.

➤ My associate John Schwartz, award-winning architect, who was the first to join me in applying this process and who has been the greatest friend and cheerleader.

➤ Kerry Nicholson and Dean Henry, successful developers and supportive clients, for the inspiration to write about this subject.

➤ Gerry Gerron, one of the world's great architects, who took precious time to read the draft, offered significant insight, and encouraged me to stick it out.

➤ Roger Williams, FAIA, architect, educator, and consummate professional, who devoted valuable effort in advising me on the early drafts.

➤ My dad, Tony, who never stopped believing in me.

➤ Heather, who lovingly nursed me through the writing process, editing 12 versions, and whose patience with this "just get it done" guy was tried and tested but never failed.

➤ Dean Henry and Kerry Nicholson, Legacy Partners Residential.

➤ Leslie Moldow, Mithun, Architect.

➤ Dan Chandler, Olympic Associates Company.

➤ Michael Gruber, for his creative contribution and for his steadfast friendship.

➤ Perkowtiz + Ruth Architects, for the Project Narrative included in the sample RFP found in the Appendix.

➤ And finally, all those I've been so fortunate to work with in this dynamic industry, who share the pride and sense of accomplishment, and without whom I would have missed this terrific ride.

Introduction

This book presents a step-by-step process for executing a Request for Proposal (RFP) for negotiated contracts. Sample formats and spreadsheets are included to illustrate various aspects of the process and to enhance the reader's understanding, but they are not intended to be definitive. Since each RFP plan varies according to the team's objectives and other relevant process-related issues, formats must be tailored accordingly.

A well-managed RFP process will lead to a successful negotiation and Guaranteed Maximum Price (GMP) contract with a qualified contractor. The goal is to select the best candidate, capable of (1) working with the team to ensure that design meets budget and (2) constructing a quality product, on time and in budget.

When the process presented in this book is utilized in a collaborative manner involving the owner, architect, and/or construction manager, it will provide a clear and defined basis for a value-oriented RFP plan, and increase the probability of achieving the greatest possible success on the project.

Successful RFPs in Construction

MANAGING THE REQUEST FOR PROPOSAL PROCESS

CHAPTER 1

Why Negotiated?

The negotiated contract approach to construction has increasingly become the delivery method of choice for owners, including private developers and even some public and semipublic partnership entities. The negotiated approach provides many benefits to the design and construction process, thereby increasing the opportunity for value-added project delivery from start to finish. Benefits include

➤ Assembling a team of qualified professionals to design and build the project in keeping with the owner's goals

➤ Creating a basis for understanding the true cost-benefit elements of critical design decisions required early in the design process

➤ Testing the design for cost prior to expending significant capital on design, thereby reducing the potential for costly redesign

➤ Affording the contractor time and access to the team to plan the construction, value engineering, and establish a strong and trusting team relationship

➤ Providing the opportunity for timely schedule feedback necessary for cost-of-carry analysis and occupancy planning

➤ Greatly reducing the possibility of adversarial and contentious relationships often encountered in the competitive-bid approach

➤ Emphasizing quality, schedule, and program as equally important as cost

➤ Selecting a contractor on the basis of proven experience, qualified personnel, and track record for success as well as cost competitiveness

The Request for Proposal (RFP) serves as the vehicle for defining the terms underpinning the negotiation and should be designed to ensure the successful execution of a negotiated Guaranteed Maximum Price (GMP) contract. Construction price typically represents two-thirds of the total project cost, making the selection of a qualified contractor—and basing that selection on clearly defined terms—one of the most important elements of the project.

The RFP provides an opportunity to fully define the project and prescribe the basis for the construction price and schedule at the early stages of design. It will serve as the foundation for the agreement between owner and contractor and should be carefully planned and managed.

The industry-standard negotiated construction contract is AIA Document A111™: *Standard Form of Agreement between Owner and Contractor—Cost of the Work Plus a Fee with a Negotiated Guaranteed Maximum Price.* This contract is utilized to specify the scope of the work and to detail certain costs, such as contractor's fee, insurance, taxes, bond, labor rates (with burden), and other elements of the total cost of performing the work. These are the costs to be negotiated prior to execution of the contract. The RFP is the document used to invite contractors to propose these costs and relevant commitments, including schedule, delivery approach, and personnel.

The negotiated Guaranteed Maximum Price is intended to reflect the maximum price of the scope of work defined in the contract documents. Many owners believe that the maximum price includes everything required to complete the project ("all-in"), regardless of the clarity of the design documents. This is especially true when the GMP is requested prior to full completion of the design documents. Incomplete scope definition often leads to disagreements about cost responsibility. If the contractor is directed to include costs to cover future scope clarification, the amount allotted may be speculative and can often be higher than necessary to account for eventualities. When numerous scopes of work are involved, the cumulative effect can be substantial. As a result, when the owner demands an "all-in" GMP, the price will likely be inflated.

The subsequent impact on the project budget will be the unnecessary inflation of construction costs, which can reduce opportunities for value-added decisions in the early stages of the project. It is better to identify the excess funds during the design stages and use them to make value-added improvements than to have them returned as savings at completion, when it is too late to make a difference.

The negotiated approach allows the formation of a team early in the design process that can give meaningful cost and schedule feedback, thereby clarifying the design to support the intended budget. A qualified contractor can participate with the design team in carefully defining the scope of work concurrently with the development of the documents. The intended result is an accurate Guaranteed Maximum Price that supports the budget, the program requirements, and the desired level of quality. Figure 1-1 presents a flowchart of such a design/contractor interface.

The contractor can also provide timely cost-benefit operational analyses, supplying the team with relevant data for design decisions. This process includes value engineering as well as comparing alternatives such as structural systems (e.g., concrete versus steel), energy-efficient mechanical systems, and exterior enclosure materials (e.g., metal panels versus stucco).

FIGURE 1-1 Preconstruction Team Budgeting Process

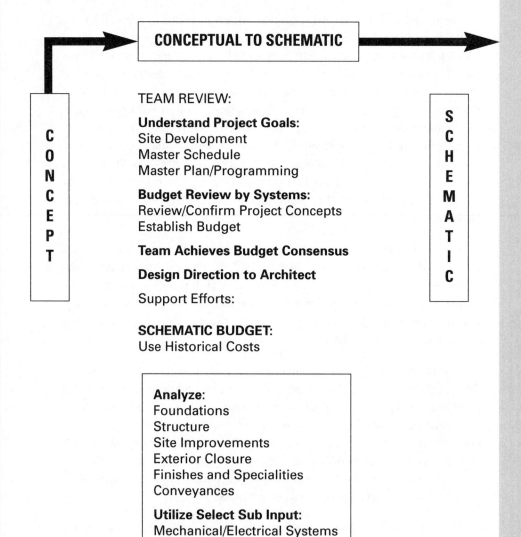

CONCEPTUAL TO SCHEMATIC

C
O
N
C
E
P
T

S
C
H
E
M
A
T
I
C

TEAM REVIEW:

Understand Project Goals:
Site Development
Master Schedule
Master Plan/Programming

Budget Review by Systems:
Review/Confirm Project Concepts
Establish Budget

Team Achieves Budget Consensus

Design Direction to Architect

Support Efforts:

SCHEMATIC BUDGET:
Use Historical Costs

Analyze:
Foundations
Structure
Site Improvements
Exterior Closure
Finishes and Specialities
Conveyances

Utilize Select Sub Input:
Mechanical/Electrical Systems
Sitework
Finishes and Specialities
Conveyances

FIGURE 1-1 (CONTINUED)

SCHEMATIC TO DESIGN DEVELOPEMENT

S C H E M A T I C

D E S I G N D E V E L O P M E N T

TEAM REVIEW:

Budget Review in Detail:
Cost vs. Function (VALUE)
Review Scope Definitions
Determine Cost/Design Studies

Design Direction to Architect

Support Efforts:

DESIGN DEVELOPMENT BUDGET:
Use Unit Cost Pricing

Analyze:
Site Finishes
Interior Finishes
Structure
Exterior Wall
Roofing

**Update Mechanical/
Electrical Systems**

Utilize Detailed Sub Input:
To Fine-tune Unit Pricing
To Develop CE* Options

Prepare CE* Options Log

*Cost Engineering

FIGURE 1-1 (CONTINUED)

DESIGN DEVELOPMENT TO CONTRACT DOCUMENTS

D E S I G N D E V E L O P M E N T

G U A R A N T E E D M A X I M U M P R I C E

TEAM REVIEW:

Budget Review in Detail:
Confirm Budget/Design "On Track"
Review and Select CE* Options
Determine Further Studies

Design Direction to Architect

Support Efforts:

DEVELOP FINAL GMP:
Use Bid Prices

> Drawing Review
> Write Bid Instructions
> Invite Sub Bidder Market
> Solicit Sub/Bidder CE*
> and Alternatives
> Keep Market Informed
> Coordinate Addenda
> Answer Bidder Questions

Update CE* Options Log to Reflect Final GMP

*Cost Engineering

Figure 1-2 represents the key elements of a successful project. Each is interrelated, and it is rarely possible to make revisions to one without impacting one or both of the others. The goal of the negotiated approach is to balance the elements of the triangle to ensure that all three project requirements are met. Cost is determined concurrent with design maturation, ensuring that timely decisions about quality and program are analyzed within the context of known cost. The architect, owner, and contractor work as a team during preconstruction to measure and balance the three elements. Expectations at the start of construction are therefore programmed into the documents and the Guaranteed Maximum Price.

In a competitive-bid approach, the construction documents are prepared to meet the constraints of the program and to set the level of quality, within the intended range of cost. Cost, however, is not verified until bids are

FIGURE 1-2

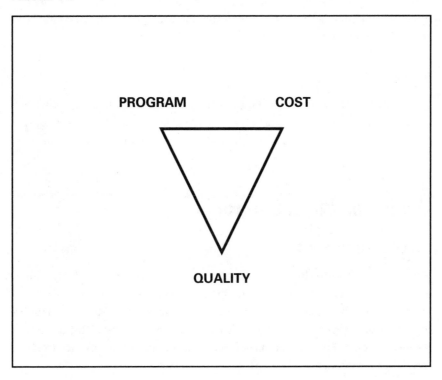

received following completion of the documents. When the bids are received *after* the documents are completed, the bid amount may exceed the requirements of the budget. Rebidding can be costly in terms of delays and dollars. Furthermore, the low-bid contractor's primary goal is to make a profit, making quality secondary.

Following is a discussion of the key differences between the negotiated "team" approach and the competitive-bid approach.

Contractor Selection

NEGOTIATED APPROACH

This approach provides for the selection of a contractor based on qualifications of the company and personnel assigned to the project, experience in similar product types, reputation, and financial stability. Key financial elements of the construction cost are negotiated, and the remaining costs are bid in the subcontractor market, ensuring that most of the construction cost is competitively bid.

BID APPROACH

The contractor is selected based on lowest cost. The contractor may have little relevant experience, unqualified personnel, and a price that neglects quality, subjecting the owner to a high degree of risk in ensuring a successful outcome.

Quality and Cost Control

NEGOTIATED APPROACH

This method provides for a team approach between contractor, design team, and owner to balancing the cost, quality, and program elements. The interactive process begins early in the design stage and affords the opportunity to assess the tradeoffs and ensure that the design tracks with the budget requirements. This approach provides a higher degree of assurance that the overall project goals will be met.

BID APPROACH

This approach requires bidders to utilize the cheapest subcontractor bids to win the job. In this manner the owner and design team are subject to multiple subcontractors whose only objective is to make a profit. Quality suffers, and when the owner and architect enforce quality standards, adversarial relationships can develop, distracting the team from the ultimate success of the project.

Value-Added Participation

NEGOTIATED APPROACH

This approach provides for selection of a contractor team with proven experience in cost-estimating design alternatives, value engineering, constructability reviews, and ensuring that subcontractor bidders are qualified and reliable. This scenario provides for cost accountability from the contractor and quantitative assessment of key design elements, thereby ensuring that program and operational goals are met. In the negotiated approach the contractor tends to assume ownership of the design and related costs.

BID APPROACH

This method provides no opportunity for value-added participation by the contractor during the design stage. The contractor assumes no ownership of design and related costs and is free to find fault in the design in order to create opportunities for change orders.

Beneficial Relationships

NEGOTIATED APPROACH

This approach provides for mutually beneficial negotiation between owner and contractor. Participants in the negotiation must come to agreeable terms to make the deal, thereby affording an opportunity for all parties to buy in. During the negotiated RFP process, team interaction, relationships with the design group, construction manager, and owner may be assessed prior to contractor selection. The owner has the opportunity to ensure that the contractor accepts the project goals for cost, quality, and program and is suited to work with the team in achieving the goals.

BID APPROACH

This approach provides no opportunity to ensure proactive working relationships and buy-in of the project goals. The goals of the contractor are in many cases contrary to those of the owner and to the design team goals, and there exists no proven relationship with the team to mediate the differences. The contractor takes no ownership of cost, quality, or program.

Cost Increases

NEGOTIATED APPROACH

This approach provides the opportunity during design for the contractor to assign cost, participate in constructability and design decisions, and manage construction to those decisions. Since the contractor updates the budget periodically as the design matures, the final GMP represents costs established by the contractor and upon which the final design is based. The contractor is more likely to "own" the cost, thereby reducing the risk of excessive cost overruns.

BID APPROACH

This method provides for cost ownership by the contractor on a contractual basis only. The cost is based on low subcontractor bids and completed bid documents. The contractor may be motivated to find "holes" in the documents in order to create opportunities to enhance profit, thereby increasing the likelihood of change orders.

Final Costs/Best Value

NEGOTIATED APPROACH

This approach can be structured to allow verification of final costs through audit or other accounting procedures. The owner pays for the actual cost of the work plus a negotiated fee and is able to verify the costs. The owner has the opportunity to recoup cost underruns.

BID APPROACH

No method is provided to verify final costs. The contractor is motivated to reduce cost in order to enhance profit, thereby compromising quality and long term durability of the final product. The owner may end up paying more for a compromised product.

Summary

The negotiated approach relies heavily on teamwork. Including the architect and engineers in the RFP process will make the effort collaborative and team-building, while enhancing the likelihood of selecting *the most qualified contractor at a fair price.* Useful methods of identifying team players are discussed in subsequent chapters. Without team players, the negotiated approach is "guaranteed" to fall short of adding "maximum" value to the design and construction process at a reasonable "price."

The basic elements of the RFP plan outlined in this book can be applied to projects of all types and sizes, ranging from tenant improvements to renovation and new construction. The process can be tailored to reflect a project's specific requirements and performance standards (e.g., schedule, cost reporting). The size of the project generally (but not always) dictates the complexity and time constraints of the RFP effort. Whatever the size and type of project, however, it is always true that the more specific the information provided in the RFP, the more predictable the outcome: meaningful cost and schedule feedback, and reliable data on which to base the contractor selection.

The subsequent chapters address the essential steps in the RFP process:

➤ Chapter 2, *The Cost:* The development team must consider and account for the cost of a thorough RFP process.

➤ Chapter 3, *The Search:* Identifying qualified contractors to receive the RFP and compete for the project is an essential part of maximizing the benefits of the RFP process.

➤ Chapter 4, *The Request for Proposal:* Composing and packaging the RFP to provide relevant project data to the candidates, and to set the stage for meaningful responses, will ensure the data are complete and organized, thereby allowing efficient and timely analysis.

➤ Chapter 5, *The Analysis:* The data provided by the contractors relate to the pro forma program, schedule, quality, and cost. Detailed and objective analysis ensures the greatest benefit from the RFP process.

➤ Chapter 6, *The Interview:* It is the people who will make the difference in the success of the project. Meeting and interacting with the team provides an opportunity to assess the fit.

➤ Chapter 7, *The Negotiation:* Tangible and intangible elements of the deal points are all subject to negotiation. It is the final opportunity to make the best deal, relating to not just cost but also personnel, schedule, and performance standards.

➤ Chapter 8, *The Deal:* Consummating the deal by enumerating all critical contract terms eliminates disagreements and misunderstandings and clears the way for managing the design to fit the budget.

➤ Chapter 9, *What's Next?:* Setting the stage for the contractor to get up to speed quickly and efficiently, and to participate in supporting critical design/cost decisions, ensures that the team is equipped to succeed.

In the ensuing chapters, examples of documents developed for use on actual projects are given to illustrate the process. A Request for Proposal developed for use on an actual project is included in the Appendix and is referenced throughout the text.

CHAPTER 2

The Cost

Every component of the development continuum has a cost, and the RFP process is no exception. Various factors affect the cost of an effective RFP plan. Some factors are consistent, regardless of project size; others are specific to project size; still others are exclusive to project type (e.g., retail, office). Other influences include the extent of experience the team leaders have with the RFP process, the quality and quantity of design documents, the owner/user project requirements, the extent of information requested in the RFP, and the overall RFP plan itself.

It is clear, then, that cost should be a consideration in formulating the RFP plan. In general, smaller projects (under $1,000,000) do not allow significant opportunity to recover sunk RFP process costs. Careful thought, therefore, must be given to the extent of the plan. An effective RFP on smaller projects need not be elaborate and costly to lead to a successful outcome. Limiting consideration to the critical elements that lead to selection of a qualified contractor and to primary deal points will result in a simplified RFP, targeting relevant information. Chapter 4, "The Request for Proposal," presents the critical requests for information to be included in the RFP and identifies those recommended for use on small projects.

The benefit of a negotiated approach is the opportunity to use the contractor's expertise in developing design documents that incorporate cost-efficient construction techniques and value-added materials while minimizing change-order risks. On larger projects, the savings generated using the negotiated approach (making cost feedback, value engineering, and schedule input available early in the design process) will likely exceed the costs of administering an effective RFP plan. It is reasonable to anticipate savings of 5 percent or more of the construction costs by using the negotiated approach, which would represent $100,000 savings on a $2,000,000 project, an amount that far exceeds the cost of administering the RFP plan.

Assuming the RFP process leads to the selection of a competent, qualified contractor experienced in working with a design team to identify cost-saving opportunities, the decision to commit dollars to a well-managed RFP process will pay off handsomely. However, administering a minimal effort in order to save money, using incomplete documents, and choosing a contractor in the absence of a detailed assessment all add significant risks to the desired successful outcome.

Following is a discussion of the key elements of the RFP cost, including a look at dollars, resources, and time.

Managing the RFP Process

The cost of managing the process is affected by numerous factors. Compensation of the manager and time contributed to the effort are obvious examples. Other costs may include travel, printing, shipping, communications, and support staff (clerical, accounting). The planned timeline for the RFP process will allow an accurate assessment of these costs in terms of dollars.

There are other costs that not only relate to dollars but also affect the outcome. Consideration must be given to any potential delay to the design, permitting, and construction start dates, all of which may be impacted by the plan. In most cases, the RFP process can be dovetailed into the ongoing design and permitting continuum without impact, provided the timing of all these elements is assessed in planning for the RFP.

For instance, administering an RFP plan near the end of the design development effort may necessitate a cessation or slowing of design, in order to allow ample opportunity for cost-saving decisions to be incorporated into the design documents without excessive additional costs. This can be avoided if the team has discussed the process, developed a plan, and fully anticipated the effects.

Permit submission may also be delayed or impacted as a result of a poorly planned process. In the event a contractor joins the team late in the design development timeline, any value-added design modification may necessitate resubmission or revision of permit documents. For example, a $100,000 cost-savings option is one an owner may not want to miss. However, if the permit documents require revision—and this causes a delay to the permit submission and subsequently the permit issuance—the cost of the resultant delayed construction start in terms of "cost of carry" (financing costs) and lost revenue at occupancy could likely far exceed the savings.

RFP-related Design Costs

RFP plans can generally be categorized as follows:

FEE-BASED APPROACH

The RFP requests that the contractor propose fees for profit, overhead, taxes, insurance, and related "add-on" costs, all applied to the actual hard cost of construction (e.g., steel, concrete, labor). Under this method, the extent of design and other project-related documents to be included in the RFP is greatly reduced. Basic conceptual floor plans, site plan, elevation, and program requirements are all that is necessary to elicit reliable proposals. (Note, however, that more information is better in reducing the possibility for misunderstandings.)

ESTIMATE-BASED APPROACH (PREFERRED METHOD)

In this case the RFP requests a detailed estimate of construction cost in addition to the fees and markups discussed above. Since this is an ideal opportunity to solicit cost input from multiple contractors, it provides a broad-based method of testing the pro forma assumptions for cost and schedule. Following receipt of the proposals and estimates, the analysis and follow-up phases of the plan provide the ideal opportunity to explore creative alternatives to design and material decisions by accessing the databases and wealth of experience available from the contractor candidates. The RFP process is most likely the only time in the project continuum when multiple contractor resources are available.

To receive reliable cost estimates, extensive design, engineering, and program information must be included in the RFP package. Generally, an RFP plan that requests a construction cost estimate from the contractors will be more costly than a plan requesting a fee proposal without estimate. Obviously, the extent of design maturation required for an estimate-based RFP plan is considerably greater than that for a fee-based plan.

Design costs associated with the RFP process are directly related to two factors: (1) the extent of design documents to be included in the RFP and (2) the timing of the RFP as it relates to the design timeline. In the event the RFP is estimate-based and occurs early in the design timelines (e.g., conceptual/schematic documents), the design team will most likely expend resources, some out of sequence, in developing the detail required to obtain meaningful estimates in the contractors' RFP responses. The example RFP included in the Appendix was produced at the conceptual design stage. Many hours and resources were expended defining the various key elements of cost for civil, structural, architectural, and other related project components.

The cost of such an effort on a large project should not be overlooked in the pro forma. The benefits, however, are likely to be well worth the cost:

> ➤ A clear and concise design program is developed at the outset and serves as the basis for the design effort, thereby enhancing efficiency in design maturation.

➤ A detailed project definition is produced and available for use in the financing package.

➤ A defined basis for construction cost is provided and is invaluable as a benchmark against which to measure value-added choices.

➤ The quality and detail included in the contractor estimates give the team a basis for assessment of contractor capabilities.

➤ The program requirements are available for owner/user review and contribute to mitigating the risk of design revisions later in the process.

These benefits may not be measurable in dollars and cents but will clearly improve the chances of a successful project.

When an estimate-based RFP plan is administered later in the design time-line, the cost of producing documents is reduced since most documents have already been developed as part of the schematic to design development continuum. The costs of such document production may be largely included in the design fees, especially if the initial design contract anticipates the RFP process.

As previously noted, however, while the cost of documentation for the RFP may be reduced as the project progresses along the design timeline, the risk of minimizing cost-saving opportunities increases, subject to the impact of design revisions, which in and of themselves may represent additional costs.

Figure 2-1 illustrates the cost-benefit relationship between design maturation and contractor participation in a negotiated approach. The team should discuss the cost-benefit elements of timing the RFP process with the design maturation. In every project, there is an optimal time to execute the process. Provided the decision is consensus-driven and incorporated into the overall design plan, benefits will almost always outweigh costs.

FIGURE 2-1

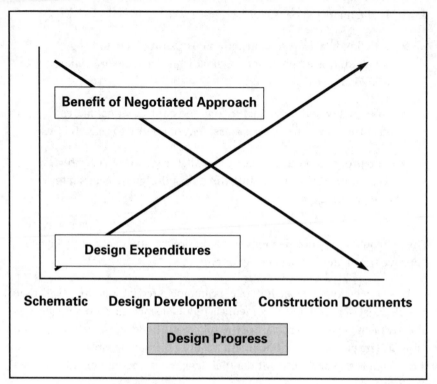

Contractor Costs

The cost to the contractor of preparing a detailed estimate and RFP response package can be significant. On large and complex projects with an extensive RFP, the cost can easily exceed $10,000 and in some cases can be as much as $20,000 or more. If there are six prequalified contractors selected to receive the RFP, the collective cost of $60,000 or more is obviously significant. While this may not be a cost to the project, it is representative of the extent of effort required to respond to a detailed RFP. This magnitude of effort by the contractors illustrates the extent of valuable information the team can obtain through a well-managed RFP plan. Given this viewpoint, imagine the resources available to the benefit of the project at no cost to the owner! This adds perspective to the benefits of a negotiated RFP approach.

Considering the cost to each contractor in competing for selection, it is of utmost importance that the process be fair and balanced and preclude favoritism. Furthermore, appropriate resources should be applied in assessing the responses, clarifying ambiguity, and ultimately making a qualified selection.

The chances of making the best selection will thereby be increased, and the quality and quantity of relevant information gathered will be considerable. Historical cost data, value engineering ideas, creative construction techniques, and opportunities for schedule enhancement are all potential points of information that can be obtained from the collective contractor responses.

Construction Costs

On the surface it may seem incongruous to suggest construction costs may be related to the RFP process, since construction occurs long after the RFP. The construction costs, however, are clearly related to the outcome of the RFP plan.

Contractor's fee, markups, labor rates, and other elements of cost will be defined as the basis of the final negotiation and deal, which culminate the RFP process. These could comprise 10 percent or more of the construction costs, representing a significant impact directly related to the RFP.

Other project costs may be impacted as an outcome of the negotiated RFP plan. Value engineering, system selection (first cost versus operational cost), and other advantages are available as a result of having a contractor join the team during the design effort. These activities represent construction cost-saving opportunities and value-added decision making that is directly related to the RFP process. These benefits should not be overlooked in assessing the cost of the RFP plan.

The Cost of Adversity

When the negotiated RFP versus the competitive-bid approach is assessed, consideration must be given to the benefits of early inclusion of a contractor on the team. A discussion of the benefits of the negotiated process was presented in Chapter 1, illustrating the various attributes. By its nature, the competitive-bid process produces potentially opposing objectives between contractor, designer, and owner/user.

To win a bid competition, the contractor must present the lowest bid, which may provide little or no margin for profit. The contractor's likely intention, then, will be to take full advantage of every design conflict to create change orders leading to enhanced profit. In other words, the contractor will be looking for ways to add costs. This generally leads to contentious relationships among the parties, each defending his or her position.

A negotiated process provides for careful selection of a contractor with a proven teamwork philosophy, a reputation for quality, and experience in creating value for the benefit of the project. Once the contractor is selected and joins the team to create the construction plan, there is opportunity and potential to develop mutually beneficial goals, team camaraderie, and effective communication channels.

As the design matures, with the contractor contributing to the process, the team can utilize this combined expertise to identify design conflicts, establish reasonable budgets, and establish clearly defined quality standards. During construction, when design conflicts arise, as they invariably do, the resolution will more likely involve a consensus approach, resulting in win/win outcomes. Integrating the three key elements of a successful project—cost, quality, and program—requires fairness, teamwork, and expertise, all of which are likely to be applied effectively in the negotiated approach. The cost of not achieving the triad can be significant. In the instance of compromised quality of the finished project, the cost may not be recoverable.

In summary, the negotiated RFP process includes costs to the pro forma, schedule, and resources that a competitive-bid approach does not. The decision to use either the negotiated or the competitive-bid process will most likely be based on considerations other than cost. Even if the cost of administering a negotiated RFP exceeds $50,000—which could be likely on large and complex projects—as a percentage of the overall pro forma, the cost is minimal.

Cost surprises in the pro forma, however, are never desirable. It is good practice to carefully assess the cost-benefit elements of the negotiated RFP process, including (1) the timing of the RFP along the design continuum, (2) the extent of the RFP package, (3) the selection criteria in determining information requested from the contractors, (4) discussion with the team to achieve consensus, and (5) timely execution as planned.

The best RFP plan for your project may or may not be complex; but in either case, if it is thoughtfully planned and carefully executed in a consensus-driven environment, cost risk will be reduced and the chances of success increased.

CHAPTER 3

The Search

Many resources are available to facilitate the search for contractors. The goal is to assemble a list of qualified and competent contractors experienced in adding value to the design process. Contractors with experience in preconstruction are essential to a successful negotiated GMP approach. The search should be designed to identify these types of contractors and to sort out traditional competitive-bid contractors who lack preconstruction expertise.

Contractors tend to have areas of expertise according to project type and size. Some may specialize in medical facilities, others in office buildings; some focus on projects under $10 million, and others on projects over $10 million. Develop your list from contractors with experience in your project type and size.

The design team may be helpful in identifying candidates. Architects and engineers have first-hand experience with contractors located within their region. Obtaining referrals from owners who have recent comparable project experience will also assist in your search.

The following criteria are useful in your query when you consult the available resources knowledgeable about qualified contractors in your area:

➤ Determine whether experience with the contractor is direct or indirect.

➤ Define the contractual relationship with the contractor.

➤ Discuss whether the experience with the contractor is recent.

➤ Determine specific project-related results: on-time, in-budget, quality.

➤ Assess if the contractor is a team player.

➤ Request identification of the strengths and weaknesses of the contractor during both the preconstruction and construction process.

➤ Review the contractor's cost management and reporting expertise.

➤ Inquire as to the quality of documentation and communication.

➤ Determine the effectiveness of the contractor's dispute resolution approach.

➤ Discuss the contractor's management style: can-do, responsive, proactive versus reactive.

➤ Determine the level of executive support and follow-up.

Discussion of these criteria will provide a better understanding of the contractor candidates' overall capabilities as well as proven track record. Successful contractors with extensive negotiated construction expertise will undoubtedly be named in the discussion by more than one resource, and the identities of the best candidates will become apparent.

Both the Associated Builders and Contractors (ABC) and the Associated General Contractors of America (AGC) maintain a computerized database on their member contractors. Both offer referral services whereby you can submit project requirements and, in most cases, receive a list of suitable contractors.

ABC
1300 N. Seventeenth Street, Suite 800
Rosslyn, VA 22209
(703) 812-2000
www.ABC.org

AGC
333 John Carlyle Street, Suite 200
Alexandria, VA 22314
(202) 548-3118
www.AGC.org

Qualified and experienced contractors tend to be members of professional trade associations, such as the ABC and the AGC, or of local organizations such as the chamber of commerce. Contractors without such affiliations are not necessarily unqualified. Membership is, however, a good indicator of a contractor's active participation in the industry and community; it can also be an indicator of responsibility and ethics.

In refining the list of qualified contractors, it is useful to request, for submission by the contractors, Form AIA A305™, *Contractor's Qualification Statement.* You may obtain this form from the architect, purchase it at the local AIA store, or order it online at www.AIA.org. Most contractors have a database relative to this form, which covers company history, licensing, general experience, annual volume, references, and bonding limits and capacity.

Keep in mind that each contractor's capability *and* workload are of equal importance. A contractor may be capable according to *resume* but incapable according to *backlog.* Analyze potential candidates based on capacity as well as qualifications.

A starting list of at least five qualified contractors is suggested. If one or two firms withdraw during the process, enough will remain to maintain a competitive effort. (A shorter list can be developed prior to the interview, after analysis of the RFP responses.) In addition, less than five RFP responses may not provide adequate information to assess each contractor's qualifications regarding the nature of the project and the nuances of the locale. Comparing five or more responses can offer a revealing look at each contractor's familiarity with, and *desire for,* your project. The teamwork approach demands a contractor who *knows* your project and *wants* to build it.

A word of caution is in order regarding the number of contractors to receive the RFP. In the event the RFP solicits extensive and detailed information from the contractors, too many responses can make the process cumbersome and costly. An excessive number of responses do not necessarily add value to the process. It is more important to be dutiful in prequalifying and determining a select list of recipients than to include too many candidates. The number of contractors on your list should reflect a meaningful cross section of proven experience, cost conscious project management, and quality construction, but limited to a manageable total.

CHAPTER 4

The Request for Proposal

The Request for Proposal presents an ideal opportunity to clearly define the scope of the project for all parties, including the design consultants. By developing a comprehensive project summary for issuance with the RFP, you will create a package that can be used in the equity/finance effort as well as in market studies and related activities. An early, refined definition of the project scope provides a basis for both the pro forma and the negotiated construction contract—and a blueprint for managing the design to meet the budget.

A generic RFP, developed for use on an actual project, is included in the Appendix for illustrative use with this book. It contains a detailed Project Narrative, which defines the scope of the project at the early design stages. An outline specification is often used in lieu of the Narrative. In either case, sufficient detail must be included to support meaningful and reliable responses to the RFP and establish a basis for contractor selection.

The Request for Proposal can be divided into four general sections (with subsections within each) fully defining the requirements specific to the project. The following discussion of each of these elements provides oversight intended to clarify and guide the design of the RFP. Since each project is different and those differences dictate the specifics of the RFP, some elements may not be relevant for inclusion. An asterisk indicates the categories that may be considered for exclusion on smaller projects. All others should be considered essential to an effective RFP.

Section A: Overview

1. INTRODUCTION

Provide, in summary form, a general description of the RFP plan. Identify the owner, location of the project, building type, due date and location for delivery of the response, a brief overview of the RFP plan, target date for selection of finalist, and planned construction start date.

2. PROJECT DESCRIPTION

Summarize the project, identifying number of stories (above and below grade), use(s) and associated breakdown of square footage according to use (e.g., parking, office, retail), and program. This section can be kept brief, as it is intended as an introductory overview only. More detailed information will be provided in the "Pricing Narrative" and related documents section.

3. IDENTIFICATION OF OWNER'S VENDORS AND CONSULTANTS *

Many owners maintain national or regional accounts with certain vendors. Regardless of the type of vendor relationship, if you intend to assign a vendor to the contractor, name the vendor(s) and scope(s) of work, and request the contractor's *written acceptance* of the assignment in the response to the RFP. This can eliminate disagreements later in the process. It may also provide the contractor with pricing resources.

It is good practice to identify all team members. Identifying the architect and the design and engineering consultants—structural, civil, electrical, mechanical, acoustical, and others—will provide the contractor with resources for budgeting and understanding the project prior to submitting the response to the RFP. Require that questions for consultants be submitted through the manager of the RFP, to ensure consistency of information flow. Consultants' written responses to questions should be distributed to all RFP respondents to maintain a fair and competitive process. Be sure to instruct all consultants to respond to contractor inquiries *only* through the RFP manager.

In the event the contractor has a positive working relationship with members of the design team, the owner may receive the benefit of favorable fee and schedule responses. Experienced contractors recognize design consultants who provide timely and meaningful participation during construction, thereby affording the contractor the means to complete the work in an efficient and cost-effective manner.

In some instances, design-build subcontractors, such as mechanical and electrical, may be chosen prior to selection of the contractor (in lieu of traditional engineering disciplines). This allows the owner the opportunity to make informed system selections that may be required to support early design decisions in the absence of the contractor. The RFP should identify any design-build subcontractors and request the contractor's *written acknowledgment* to accept assignment of these subcontractors (if that is the intention). It is suggested that this acceptance be made a strict condition of a qualified response to the RFP.

Likewise if design-build for certain disciplines is intended to be included in the contractor's scope, identify the scope(s) and define the expectations for performance; for example,

➤ Is the permit to be included?

➤ Are the design documents to include an engineer's stamp?

➤ Are complete specifications required?

➤ Is electronic design required (e.g., CAD)?

➤ Is professional liability insurance required (specific amount)?

4. RELEVANT SCHEDULE INFORMATION

The pro forma schedule for construction may impact the fees and other costs proposed by the contractor in the RFP response. Material cost escalation, backlog, subcontractor and workforce availability, and other relevant issues affect the contractor's approach to the cost of construction and are dependent on the schedule requirements. The timing of the project is also key in the contractor's selection of personnel, including the superintendent and project manager.

Identify the following timelines in the RFP:

➤ Begin preconstruction no later than _____.

➤ Establish final GMP by _____.

➤ Begin construction on or about _____.

It is suggested that the anticipated construction start be identified by quarter rather than by date, e.g., second quarter 2006. Identifying a specific date and then missing it may provide the contractor with the opportunity to request additional compensation and/or to change personnel.

➤ Complete construction, ready for occupancy on _____.

The pro forma typically anticipates construction duration. The contractor will set pricing, fee, and general requirements accordingly. It is important to tie the contractor's response to the pro forma schedule requirements. Another option is to request duration from the contractor. When this option is chosen, it is essential to confirm that the overall duration accommodates the pro forma cost-of-carry. (The basis of completion is typically defined by the contract; reference to that definition may be made in this section.)

In the event penalties for late completion are contemplated, they should be identified with a request for *written acknowledgment*. Contractors often calculate costs and fees differently if at risk for completion penalties; the contractor's proposed schedule may be impacted as well.

Additional schedule requirements may need to be specified in the RFP. Examples include

➤ Key milestones such as phased permit submittals

➤ Budget updates

➤ Phased occupancy requirements

While it is best to be general and not to pin down exact dates, some milestones may be fixed and critical to the success of the project. These should be clearly identified and defined.

5. DELIVERY METHOD

A description of the intended method of contracting should be included indicating the type of contract to be utilized. As previously noted, the industry-standard negotiated construction contract is AIA Document A111™: *Standard Form of Agreement between Owner and Contractor—Cost of the Work Plus a Fee with a Negotiated Guaranteed Maximum Price.* This contract is recognized by the design and contractor community and is coordinated with the architect's AIA contract. Together they clearly delineate responsibility between owner, contractor, and architect. AIA contract documents are available in electronic format at www.AIA.org; they may also be obtained through the architect or purchased at the local AIA store.

In some cases, you may choose to use a hybrid or custom contract. Consider this option carefully, however, as such contracts can lead to extensive and/or contentious negotiations. Because of the contractor's unfamiliarity with the document, expect extra scrutiny to be applied. In addition, it will be important to coordinate the responsibilities of the architect, contractor, and owner throughout the contract documents, including the related design agreements and the specifications.

An alternative to using a hybrid contract is to identify revisions to the AIA document via addendum. The revisions should be specific to each section of the contract so that they may be addressed and negotiated on an itemized basis.

A directive for the contractor to note *all* exceptions *in writing* should be included. (Some contractor's exceptions, such as exclusions, may be cause for disqualification.) This will mitigate the potential for misunderstandings when the contract is fully negotiated and executed.

It is advisable to include a brief discussion reserving the owner's right to alternative delivery methods should the project be over-budget or not be constructed. More detailed termination language is generally included in the contract, and if so, it should be noted here for reference. Use caution not to contradict the contract language.

As always, it is advisable to review any contract revisions with legal counsel.

6. PROJECT MANAGEMENT AND CONTRACTOR REQUIREMENTS*

This section may be used to prescribe expectations for the contractor's execution of certain preconstruction and construction tasks. Obtaining performance commitments in the response to the RFP ensures that the preconstruction and construction process will meet expectations.

Examples of performance requirements include

➤ Regular design-review meetings

➤ Reporting methods

➤ Interfacing with a tenant build-out during construction

➤ Coordinating the installation of the owner's data management systems and FF&E (fixtures, furnishings, and equipment)

A contractor with experience in these areas would be a value-added member of your team.

It is essential to define the contractor's duties. Provide a detailed description of the tasks required to accomplish the various critical milestones including those related to design, budgeting, permitting, and financing. Examples of preconstruction activities include

- ➤ Periodic budget updates

- ➤ Value engineering sessions

- ➤ Constructability reviews

- ➤ System selection analysis (e.g., electrical, mechanical)

- ➤ Design-build subcontractor involvement

Examples of construction activities include

- ➤ Weekly site meetings and associated minutes

- ➤ Periodic inspections

- ➤ Monthly progress reporting

- ➤ Periodic progress photographs

It is advisable to define the extent of electronic project management desired. For example, the design team may intend to utilize a web-based project management system which provides for contractor participation. Contractors familiar with the system(s) are likely to indicate so in their responses, making them uniquely qualified for this aspect of the project.

A good safety performance record on the part of the contractor is of utmost importance in mitigating the owner's risk. The RFP should request the contractor's documented safety record, which should then be verified. Serious job site injuries often end up in litigation that includes the owner. This is a case of "an ounce of prevention...."

The RFP in the Appendix includes sample performance and project requirements under the heading "Project Management/Contractor Requirements," defining preconstruction and construction phase services.

7. SELECTION CRITERIA

It is helpful *and* fair to let the contractor know the basis for selection in advance. You will want to carefully determine your criteria, however, so that you do not inadvertently encourage the contractor to submit misinformation to get the job. For example, if one of the selection criteria is lowest estimate, some contractors will provide artificially low estimates that may not be reliable in validating the pro forma and may be subject to increases during the design phase.

Meaningful selection criteria can include qualified personnel, adequate resources to meet the schedule requirements, similar completed projects, financial stability, and competitive fee and markups.

8. INTERVIEW SCHEDULE*

Identify the time and place of the interview. This will allow adequate preparation time and shorten the period between receipt of the RFP responses, the interviews, and the selection. By giving the contractors advance notice and time to prepare, you lay the groundwork for a meaningful and informative interview.

Since some contractors will likely be eliminated during the RFP analysis, include a statement that contractors will be notified if an interview is deemed necessary.

Section B: Response Format and Definitions

1. FORMAT FOR THE RFP RESPONSE

A detailed RFP requires submission of extensive information from the contractor. Analysis of the information can be time-consuming and confusing if it has not been presented by all contractors in consistent format.

Be explicit in instructing the contractor regarding presentation of information. Also be specific about what *not* to include. Corporate brochures and other marketing pieces are typically not useful in completing the analysis. It is probable that you will have already reviewed such information in establishing your list of qualified RFP recipients. You may want to direct that the contractor's response be brief and to the point. It is best to direct that the response follow the RFP format to ensure consistency in comparing multiple contractor responses.

2. REQUEST FOR LIST OF CONTRACTOR'S KEY PROJECT PERSONNEL

Perhaps the most decisive factor in selecting the best contractor among qualified firms is the personnel. All recipients of the RFP may be suitable contractors, capable of constructing the project. What will make the difference between merely completing the project and completing the project with excellence is the people—including, and perhaps most importantly, the superintendent and the project manager.

The RFP should request resumes and references for the key project personnel. The contractor's *written commitment* to provide these personnel throughout preconstruction and construction is imperative, to avoid the bait-and-switch trap.

3. REQUEST FOR KEY FINANCIAL ELEMENTS OF THE PROPOSAL

The RFP in the Appendix identifies numerous items to be included under this heading. (Refer to Section B, "Response Format and Definitions," item 3.) Award of a negotiated contract is usually determined by fee, markups, labor rates (and burden), general requirements, and qualifications (including personnel), regardless of the RFP approach: fee-based or estimate-based.

The RFP should request that the contractor state the proposed fee as a percentage of the total construction costs or as a lump sum. If your intent is to convert the percent fee to a lump-sum amount at completion of the design documents and final GMP, it may be specifically noted; or you may want to defer the lump-sum fee discussion until the final negotiation (see Chapter 7). Clear definition of the fee, overhead, and profit and how they are calculated is required to assess each contractor's response. For example, the fee may *include* overhead costs, or overhead costs may be an *additional* add-on.

Contractors often include additional markups on the subtotal of construction costs, such as applicable taxes, insurance, and contingency. The RFP must request specificity for all markups and clarification of the calculation. (For example, does the fee multiplier apply to the other markups, such as insurance and taxes, or just to the subtotal of construction costs?)

Labor rates for all trades employed by the contractor should also be requested. It is important to establish the rates to be used during preconstruction and construction, as well as in change-order pricing. Labor rates differ among contractors and, depending on the amount of self-performed work, can significantly affect the construction cost.

Labor burden is a contractor markup that is often overlooked in the cost analysis. Labor burden consists of contractor costs related to the employees' base wage rates or salaries. These include employer's share of Social Security and Medicare, federal unemployment tax (FUTA), state unemployment tax (SUTA), and other regulated expenses on wages. In addition, unions collect fees on labor rates for contributions to pension funds, health insurance, training, education, vacation funds, etc. Taken together, these costs represent the "burden" on wage rates.

Contractors calculate burden rates by using various methods. Burden may be charged as "cost of the work," based on actual costs, or as a percentage multiplier to be applied against the stated wage rates. Construction contracts often fail to address burden rates, how they are to be calculated, and how they are to be accounted and charged against the cost of the work. While one contractor may include burden in the stated wage rates, another may calculate it as an add-on using actual costs, and still another by using a percentage multiplier; there is no consistent method in the industry.

Thus, it is imperative to request in the RFP the proposed burden rates, the method of calculation, and specific backup, enumerating each component of the burden rate. Labor burden rates vary from contractor to contractor and will affect the overall cost of the project. For example, union carpenter labor rates are the same for each contractor, but the burden rates may not be the same. Contractor A's method of burden calculation may present a 50 percent burden rate, while contractor B's method may present 40 percent. If both contractors self-perform $1 million of work, the net difference in cost to the project is *$100,000!*

This financial section should also be used to address return of construction cost savings (if any) to the owner. A negotiated approach to construction affords the owner the opportunity to audit final construction costs. In the event that the actual cost is less than the GMP, the disposition of the savings must be addressed in the contract. Use this section either to specify how the savings will be split or returned or to request that a creative savings proposal from the contractor be included in the response.

It is important to determine the contractor's ability to deliver the work given the current backlog. In addition, assessing the contractor's capabilities to construct similar-size projects concurrent with other ongoing projects can be a revealing indicator of capacity and capability. Request a list of the contractor's projects including dollar value over the past 3 to 5 years, organized annually. You will then be able to determine if your project is within the contractor's historical capacity to perform.

4. REQUEST FOR CONTRACTOR'S BONDING CAPACITY, INSURANCE LIMITS, AND FINANCIALS

Bonds

The decision to bond or not to bond is a controversial one. Many owners feel the cost of a bond is excessive and unnecessary. If, prior to contractor selection, appropriate scrutiny is applied to the candidate's financial condition and track record, the risk of default can be measured and a bond may represent an unnecessary cost.

Contractors' bonding rates and capacities differ according to annual volume, backlog, balance sheet, accounts receivable, and payment history. Most contractors have a maximum bonding limit. One large job may exceed the limit—either alone or together with other jobs. Generally, the lower the bonding rates, the healthier the contractor's financial condition.

Other variables to consider include the quality of the bonding company (surety) and the contractor's history relating to bond claims. The RFP should request a letter from the contractor's surety, indicating

➤ Bonding rates

➤ Acknowledgment to provide a bond for the anticipated contract amount

➤ The name and rating of the bonding company

➤ History of claims against contractor's bond

➤ Duration of relationship between surety and contractor

The duration of the relationship between the contractor and surety is also revealing. A long-standing relationship with a high-quality surety and a positive letter from that surety are powerful indicators that a contractor has a good performance record and is financially sound.

Often the lender will require a bond as a condition of the loan, but this may not be known until later when the lender is committed. With the acknowledgment of the surety included in the contractor's response to the RFP, the owner will have relevant bonding information to provide to the lender.

If the surety has committed to bonding, the contractor is obligated to bond upon the lender's demand; however, many sureties will not unconditionally acknowledge provision of a bond without reviewing the construction contract, including the Guaranteed Maximum Price. It may be too early in the contracting process to obtain bonding commitment, but the bonding information obtained in the RFP response should provide reasonable assurance that, should one be required, a bond can and will be provided. In the rare occurrence where a contractor has been selected and subsequently cannot bond, alternative solutions are available, including

➤ Bonding the subcontractors in lieu of the general contractor

- Replacing the contractor

- Conferring with the lender to modify the bond requirements

- A cash bond

All alternatives are secondary to understanding the reasons why the contractor cannot or will not bond. In general, if a contractor cannot bond, it should be considered an indication of increased risk. In this case, strong consideration should be given to using another contractor.

Insurance

Certificates of insurance are revealing and should also be requested. Specify that such certificates indicate

- Limits of insurance and types of coverage (consult with your insurance carrier to determine minimum limits, and specify them in the RFP and/or the contract)

- Name of insurance company

- Rating of insurance company (consult with your insurance carrier to determine minimum ratings, and specify in the RFP)

- Duration of relationship between insurer and contractor

Higher limits of liability can be indicative of a strong financial condition. The insurance carrier's ratings are good indicators of a contractor's safety record and claims record, since more highly rated carriers are selective in insuring contractors with exceptional performance and claims records.

It is advisable to request a review of the insurance certificate by the owner's insurance carrier to confirm acceptable ratings and adequate limits of coverage. The project's overall insurance program must be coordinated carefully with the various contracts. Requiring the contractor to name the owner as an additional insured often adds a layer of protection at little or no additional cost. If this is desired, it should be so noted in the RFP.

Financial Statement

Include in the RFP a request for the contractor's most recent financial statement. Some contractors are reluctant to release this information with the response. In such cases, you can offer to meet with the contractor's financial officer to review the statement. The contractor's *response* to the request for the statement, in addition to the statement itself, can be a good indicator of the company's financial condition. In other words, resistance to supplying or reviewing the statement may indicate financial trouble. A willingness to supply the statement or agree to a review indicates an open-book policy, a positive indication of a responsible contractor.

5. REQUEST FOR CONTRACTOR'S WRITTEN COMMITMENT

It is important to ensure that your project will receive the appropriate commitment from the contractor to devote the resources and financial strength of the company. Until such time as a contract is executed providing a legally binding agreement to deliver the project under specific and defined terms, a letter signed by an officer of the company accompanying the RFP response will provide reassurance to the owner. Such a letter will attract the attention of the officers of the construction company and ensure they are informed about the project requirements. Request a letter on company stationery committing

- ➤ The full faith and credit of the firm in the execution of the work

- ➤ The commitment of resources and personnel required to ensure successful completion of the work

- ➤ Commitment to execute the contract with only those exceptions noted in the RFP response

6. REQUEST FOR SAMPLES OF PROJECT MANAGEMENT TOOLS*

The RFP should include a request for samples of the various key management documents utilized by the contractor. Examples include

- Change-order logs

- Submittal logs

- Field question logs

- Cost reports

- Schedule of values

- Meeting minutes

- Other relevant reports and documents

The level of sophistication of these documents will be a good indicator of the contractor's ability to efficiently manage the work.

7. REQUEST FOR CONTRACTOR'S REFERENCES*

Include in the RFP a request for references specific to similar recent projects. References should include owners, architects, facility managers, and, in some cases, major subcontractors (e.g., mechanical, electrical). Request a minimum of three of each, including company name, contact, and phone number.

For some projects, references may have been sufficiently researched during the qualification process, and this section may therefore be omitted.

References are only reliable if the follow-up is thorough. The person assigned to check references should ask appropriate questions and take detailed notes. You may want to determine the questions with members of the design team in advance. Some key topics include completion performance, quality, cooperation, working relationship with the municipality, and postcompletion follow-through.

In checking with references it is important to determine

- The relationship between the parties

- The type of project (similar or different)

- The extent of interaction with this reference

- Performance: on-time, in-budget, quality

- Willingness to use (or work with) the contractor again

- Contractor's demonstrated commitment to the overall project goals

- Contractor's interaction with other team members

- Contractor's working relationship with municipalities

- Reliability of cost estimating during preconstruction

- Proactive "can do" attitude

- Specific challenges overcome by the contractor

- Warranty management/customer service

Keep in mind that the references are provided by the contractor and are therefore likely to be positive. A good way to verify a contractor's record is to ask the references for *other contacts* with experience with that contractor—or to ask the references about their experience with the *other contractors* being considered. The local design community is also a good resource in the evaluation process.

8. REQUEST FOR PROOF OF EXPERTISE*

The RFP should request documentation supporting the contractor's expertise in the product type, including pictures of similar projects and relevant data, such as

- Name and location

- Owner (including contact information)

➤ Final contract amount

➤ Duration and completion date

➤ Gross square feet of components (e.g., garage, office, retail)

➤ Type of structure

➤ Special recognition and awards

You may have reviewed some of this information during the contractor pre-qualification process, and if so, this section may be omitted. Nonetheless, the specifics requested in this section of the RFP (e.g., duration and cost) may lend valuable information for pro forma refinement. The RFP process is the opportunity to gather relevant and meaningful cost and schedule information on similar projects that is not otherwise readily available after contractor selection.

9. REQUEST FOR IDENTIFICATION OF SELF-PERFORMED WORK*

Contractors often self-perform many of the elements of construction using their own forces. In these cases the contractor's GMP will include the cost of this self-performance, but it will not be competitively bid to the subcontractor community. The owner may not be receiving the best pricing, and in the event the scope of work is large, the impact on cost could be significant; the team will have limited recourse in verifying the self-performed pricing.

It is best to request that the contractor identify the scope(s) of work considered for self-performance. In an estimate-based RFP plan, request that the contractor not only identify but also *quantify* the cost of each scope of self-performed work. This will allow comparison of costs for self-performed scopes of work common to multiple contactors, which in turn provides an informative cost comparison process. Furthermore the analysis and follow-up may reveal contractor capabilities and expertise beneficial to the success of the project.

Self-performed work may be essential to the contractor's control of the schedule and/or quality. For instance, a contractor with a proven track record in concrete forming may be superior to the other bidders in meeting schedule and quality requirements. This may in fact be beneficial to the success of the project and therefore should not be precluded as a viable component of the project.

As another example, the quality of lobby finishes may be the key feature of the project, supporting high rental rates and successful leasing efforts. A contractor may present a highly skilled finishes crew with proven experience in that scope of work, thereby providing opportunity to ensure success of a key project objective.

By requiring the contractor to submit a bid for the self-performed scope of work directly to the owner (or architect or construction manager) along with the other bids (in that scope of work), competitive pricing can be verified.

Include a notice in the RFP reserving the right to require the contractor to bid all scopes of work (including self-performed) during assembly of the GMP. The contractor's bid for self-performed work can be compared against other bidders to ensure that you are receiving competitive pricing. In this manner the methodology for establishing a competitive GMP is established at the outset.

10. REQUEST FOR CONTRACTOR'S PROPOSED SCHEDULE

The design and owner project milestones identified in Section A ("Overview") and the project data included in Section D ("Pricing Package and Documents") provide adequate information for the contractor to assemble a reliable schedule. Request that a preconstruction and construction schedule be provided incorporating these key owner and design milestones into the contractor's plan.

It is essential to request identification of key milestones to be met by the *owner and design team* in order for the contractor to meet the proposed schedule. All long-lead orders for equipment should also be identified with deadline dates indicated as milestones. This information is essential to planning the design timeline relating to key operational and design decisions, such as mechanical system selection, structural system selection, and shoring system selection.

This is the first opportunity to test the pro forma assumptions relating to cost-of-carry (duration) and occupancy deadlines (e.g., lease commitments, marketing plan). The contractor's proposed schedule will be an important element of the final deal. It may be subject to negotiation to meet the pro forma requirements. Discussion of possible schedule tradeoffs is included in Chapter 7.

11. REQUEST FOR HISTORY OF CLAIMS AND LITIGATION

A request for the history of the firm's past and present claims and litigation should be included. Ask that the contractor identify all claims and litigation over the past 10 years and any outstanding claims and pending disputes, as well as the associated dollar amounts. You may want to specify a minimum claim amount to eliminate frivolous claims that are unlikely to affect the contractor's financial condition.

An extensive history of claims and litigation may indicate an adversarial managerial style. In addition, it may point to poor subcontractor relationships, inexperienced staff, or lack of adequate oversight by senior management. Contractors with minimal litigation and claims history are generally good managers and team players—definitely an asset to your team.

12. FORMAT FOR THE ESTIMATE *

Estimate-based Plan (Preferred)

Of the two approaches for the RFP plan (estimate-based and fee-based), the preferred method is estimate-based for the following reasons:

➤ It provides early cost feedback useful in verifying pro forma assumptions.

➤ It allows for detailed discovery during the RFP response follow-up phase, which often presents beneficial design alternatives.

➤ It establishes a baseline for managing the design to meet the budget, by establishing cost breakdowns (e.g., structure, enclosure, finishes) within which to study value-added program decisions.

➤ It allows the team access to multiple contractor cost databases for similar completed projects, based on actual local costs not otherwise available.

In the estimate-based approach, it likely is too early in the design continuum to include this preliminary estimate in the GMP contract. The documents at the early design stages are incomplete, and the budget is therefore less reliable. Such an estimate is useful, however, in validating the pro forma assumptions and establishing a detailed cost baseline to guide the team.

Every effort should be made to provide detailed scope definition to the contractors. This can include drawings, reports, surveys, and an outline specification or scope summary.

The Project Narrative in Section D of the example RFP found in the Appendix was produced by Perkowtiz and Ruth Architects for a project in Portland, Oregon. It is an excellent example of a detailed scope summary for a multi-family project. Although other project types (e.g., office, retail, hospitality) will require different scope definitions, the purpose will remain the same: accurate and detailed project definition.

Many elements may best be defined using an outline specification prepared by the architect and/or consultants for the RFP pricing effort. Early definition at this level of detail is essential to obtaining a meaningful estimate from the contractor. It also provides a road map for design and contractor progress pricing and therefore establishes a clear baseline against which to measure progress.

When a construction estimate is requested, careful analysis of the costs is of utmost importance. Figure 4-1 on pp. 48 and 49 is an example of a spreadsheet identifying the fundamental cost elements of a project using select categories from the UniFormat™. Instruct the contractors to specify pricing in an executive summary format similar to that found in the left-hand column—and to include a detailed budget backup for each heading. The spreadsheet uses a UniFormat™, but the AIA MasterFormat™ may also be used, depending on team preference. Generally, early in the design process, the UniFormat™ is best suited to analyze the cost of the overall building components (e.g., structure, enclosure), thereby supporting timely value-added design decisions.

Information on the Construction Specifications Institute (CSI) categories entitled MasterFormat™ and UniFormat™ is available for viewing online at www.csinet.org. The CSI has cross-referenced the two formats, affording the opportunity to relate the elements of cost to specific categories within either format. Be sure to assess copyright infringement considerations prior to utilizing the formats.

The categories utilized in the examples are drawn from the UniFormat™ option and simplified for illustrative purposes. The team should review the organization of categories within the desired format to determine the appropriate categories for the project.

Whether in UniFormat™ or MasterFormat™, the contractor pricing under the summary headings will simplify and expedite the analysis. Although the pricing may not be entirely reliable because of the incomplete nature of the documents, comparison of pricing from contractor to contractor can be revealing. Follow-up questioning may be required to fully understand the differences. Subsequent adjustments to the contractor's estimate may be required during the cost analysis. Follow-up and related estimate adjustments are discussed in Chapter 5.

Budget pricing in the conceptual or schematic stages can be useful in assessing the quality of information provided by the contractor and in preparing for the interview. Total project costs, however, may be subject to continuing refinement until the RFP process has reached the final stages of negotiation and the documents have matured to the design development phase. At that point, the team should have a clear understanding of the scope of work, and the contractor should be armed with adequate information to produce a refined and reliable price. In general, however, if the project documentation provided in the RFP is detailed and comprehensive, the average of all contractors' estimates provides a meaningful first validation of the pro forma assumptions.

Fee-based Plan

The fee-based approach should be utilized in the event that little or no design work has been completed and the team views the contractor as an essential element of the team to be assembled prior to developing schematic design

FIGURE 4-1 **Cost Estimate Analysis**

Total Project Gross SF	340,000	SF
Net Rentable SF	185,000	SF
Number of Units	200	EA

UPDATED PROPOSALS	CONTRACTOR A			CONTRACTOR B		
		% of			% of	
UniFormat BREAKDOWN	Total	Subtotal	$/GSF	Total	Subtotal	$/GSF
STRUCTURE		0.00%	$0.00		0.00%	$0.00
EXTERIOR CLOSURE		0.00%	$0.00		0.00%	$0.00
ROOFING		0.00%	$0.00		0.00%	$0.00
INTERIOR CONSTRUCTION		0.00%	$0.00		0.00%	$0.00
CONVEYING SYSTEMS		0.00%	$0.00		0.00%	$0.00
PLUMBING SYSTEMS		0.00%	$0.00		0.00%	$0.00
HVAC SYSTEMS		0.00%	$0.00		0.00%	$0.00
FIRE PROTECTION		0.00%	$0.00		0.00%	$0.00
ELECTRICAL SYSTEMS		0.00%	$0.00		0.00%	$0.00
SITE WORK/DEMOLITION		0.00%	$0.00		0.00%	$0.00
GENERAL REQUIREMENTS		0.00%	$0.00		0.00%	$0.00
SUBTOTAL	$0	0.00%	$0.00	$0	0.00%	$0.00
Contractor's Fee		0.00%	$0.00		0.00%	$0.00
State & Local Taxes		0.00%	$0.00		0.00%	$0.00
Insurance		0.00%	$0.00		0.00%	$0.00
Contractor's Contingency		0.00%	$0.00		0.00%	$0.00
Bond		0.00%	$0.00		0.00%	$0.00
Escalation		0.00%	$0.00		0.00%	$0.00
Preconstruction Fee		0.00%	$0.00		0.00%	$0.00
Other		0.00%	$0.00		0.00%	$0.00
TOTAL	$0		$0.00	$0		$0.00

| CONTRACTOR C | | | AVERAGE | | | NOTES |
Total	% of Subtotal	$/GSF	Total	% of Subtotal	$/GSF	
	0.00%	$0.00		0.00%	$0.00	
	0.00%	$0.00		0.00%	$0.00	
	0.00%	$0.00		0.00%	$0.00	
	0.00%	$0.00		0.00%	$0.00	
	0.00%	$0.00		0.00%	$0.00	
	0.00%	$0.00		0.00%	$0.00	
	0.00%	$0.00		0.00%	$0.00	
	0.00%	$0.00		0.00%	$0.00	
	0.00%	$0.00		0.00%	$0.00	
	0.00%	$0.00		0.00%	$0.00	
	0.00%	$0.00		0.00%	$0.00	
$0	0.00%	$0.00	$0	0.00%	$0.00	
	0.00%	$0.00		0.00%	$0.00	
	0.00%	$0.00		0.00%	$0.00	
	0.00%	$0.00		0.00%	$0.00	
	0.00%	$0.00		0.00%	$0.00	
	0.00%	$0.00		0.00%	$0.00	
	0.00%	$0.00		0.00%	$0.00	
	0.00%	$0.00		0.00%	$0.00	
	0.00%	$0.00		0.00%	$0.00	
$0		$0.00	$0		$0.00	

49

documents. This approach is less costly since design documents and narratives are not developed prior to issuance of the RFP. However, this method subjects the owner to cost risk since selection of the contractor occurs prior to establishing a baseline cost.

In this approach, the defining financial elements of the proposal include fee, markups, labor rates (and burden), and (possibly) general requirements, which are discussed under the headings "Key Financial Elements" and "Definition of Project Costs."

Whether the RFP plan is estimate-based or fee-based, these financial elements make up the consistent basis for the final deal. While a cost estimate may be too preliminary to include in the contract, these financial deal points are set and readily inserted into the agreement. Negotiating and documenting these deal points are discussed in Chapters 7 and 8.

13. DEFINITION OF PROJECT COSTS*

The RFP should define in detail the basis of project costs, including construction-related costs as well as costs of relevant permits and general requirements (GRs).

Request a point-by-point breakdown of contractor GRs. A matrix of GRs can be provided with the RFP for further specificity. You may decide to use a yes/no format to confirm inclusion by category; or, for larger projects, you may prefer a more extensive format to allow analysis of each proposal.

Figure 4-2 on pp. 52 to 57 is an example of a generic GRs analysis spreadsheet. By including such a spreadsheet with the RFP, thereby defining the GRs categories to be utilized by all respondents, you will ensure that responses are organized in consistent format. Contractors approach GRs with various job management procedures, so flexibility in the GRs responses should be accommodated. Invite the contractors to add items to ensure a complete and revealing response from all. Electronic spreadsheets are easily modified to accommodate this flexibility and allow efficient entry of the data provided in the responses. A great majority of the industry-standard GRs categories will likely be utilized by most contractors ensuring comparisons will be meaningful. If you require the contractor to use these forms, entering the data following receipt of the responses will expedite and simplify the analysis, particularly if electronic formats are provided to the contractors with the RFP.

The GRs listed in the example are generic and intended for illustration of the process. Specific GRs listings are included in the AIA MasterFormat™ (previously mentioned) under Division 01.

When the RFP contains adequate information regarding size of the project, schedule, and basis for costs, the contractor can accurately predict GRs costs. It may be advisable to fix these costs as part of the final deal. Since GRs represent a significant portion of the construction cost, fixing them in the final negotiated agreement ensures that this component of cost is no longer subject to increases (assuming the scope of work remains constant). The goal of the negotiated process is to fix costs as the design matures, thereby reducing the risk of cost surprises upon completion of the documents.

Many contractors expect compensation for preconstruction activities, such as meeting attendance, cost estimating, scheduling, and value engineering. A request to separate these costs from the construction budget should be included in the RFP. (Often contractors who do not charge for preconstruction services offer only minimal contribution, resulting in less than a "best possible outcome.")

It may be advisable to ask if the contractor will accept payment of preconstruction costs at the time of the first construction billing. (You may choose to qualify this request by agreeing to pay preconstruction costs in the event the project does not proceed to construction.) By deferring the preconstruction costs until the first billing, up-front costs can be reduced and preconstruction costs can be included as a part of the cost of the work after financing is in place.

In the estimate-based plan, the RFP should request a complete description what is included in and excluded from the project costs. This description most often comes in the form of a narrative accompanying the estimate entitled "Inclusions, Exclusions, and Qualifications." Careful scrutiny should be applied when reviewing this document to avoid overlooking costs that should be considered in the pro forma, or are "double-covered" in the pro forma *and* the estimate.

FIGURE 4-2 # General Requirements Analysis

XYZ CORPORATION BUILDING	CONTRACTOR A Total SF: 340,000		CONTRACTOR B Total SF: 340,000	
Description	$	$/SF	$	$/SF
PROJECT STAFF & SUPPORT				
Project Manager	-	0.00	-	0.00
Project Superintendent	-	0.00	-	0.00
Project Engineer	-	0.00	-	0.00
Cost/Scheduling Engineer	-	0.00	-	0.00
Office Administrator	-	0.00	-	0.00
Travel & Subsistence	-	0.00	-	0.00
Project Legal Fees	-	0.00	-	0.00
Professional Engineering Services	-	0.00	-	0.00
Scheduling	-	0.00	-	0.00
Total Project Staff & Support	-	0.00	-	0.00
PROJECT SAFETY/SECURITY				
Safety Training	-	0.00	-	0.00
Safety Engineer	-	0.00	-	0.00
Security/Watchman	-	0.00	-	0.00
Security Equipment	-	0.00	-	0.00
Total Project Safety/Security	-	0.00	-	0.00
FIELD SUPERVISION & FIELD ENGINEERING				
Field Supervision	-	0.00	-	0.00
Field Engineering	-	0.00	-	0.00
Engineering Equipment	-	0.00	-	0.00
Engineering Supplies	-	0.00	-	0.00
As-built Documents	-	0.00	-	0.00
Quality Control	-	0.00	-	0.00
Certified Survey	-	0.00	-	0.00
Total Field Supervision & Field Engineering	-	0.00	-	0.00
PROJECT OFFICE				
Office Equipment	-	0.00	-	0.00
Office Supplies	-	0.00	-	0.00
Telephone/Data	-	0.00	-	0.00
Postage & Shipping	-	0.00	-	0.00
Copies/Reproduction	-	0.00	-	0.00
Drinking Water	-	0.00	-	0.00
Project Office Setup/Dismantle	-	0.00	-	0.00

CONTRACTOR C Total SF: 340,000		NOTES
$	$/SF	
-	0.00	
-	0.00	
-	0.00	
-	0.00	
-	0.00	
-	0.00	
-	0.00	
-	0.00	
-	0.00	
-	0.00	
-	0.00	
-	0.00	
-	0.00	
-	0.00	
-	0.00	
-	0.00	
-	0.00	
-	0.00	
-	0.00	
-	0.00	
-	0.00	
-	0.00	
-	0.00	
-	0.00	
-	0.00	
-	0.00	
-	0.00	
-	0.00	
-	0.00	
-	0.00	

FIGURE 4-2 (CONTINUED)

XYZ CORPORATION BUILDING	CONTRACTOR A Total SF: 340,000		CONTRACTOR B Total SF: 340,000	
Description	$	$/SF	$	$/SF
PROJECT OFFICE (cont.)				
Project Office Rental/Maintenance	-	0.00	-	0.00
Copy Machine	-	0.00	-	0.00
Computers	-	0.00	-	0.00
Printers & Fax Machine	-	0.00	-	0.00
Furniture & Equipment	-	0.00	-	0.00
Total Project Office	-	0.00	-	0.00
TEMPORARY FACILITIES & SUPPORT				
Trailer Rental	-	0.00	-	0.00
Temporary Fence & Maintenance	-	0.00	-	0.00
Shop Costs	-	0.00	-	0.00
Tool & Dry Sheds	-	0.00	-	0.00
Project Signs	-	0.00	-	0.00
Temporary Guardrails	-	0.00	-	0.00
Safety Equipment & Supplies	-	0.00	-	0.00
Weather Protection	-	0.00	-	0.00
Temporary Stair Towers	-	0.00	-	0.00
Temporary Ladders	-	0.00	-	0.00
Temporary Fire Protection	-	0.00	-	0.00
Protect Finishes	-	0.00	-	0.00
Temporary Roads/Parking	-	0.00	-	0.00
Street Sweeping/Dust Control	-	0.00	-	0.00
Street and Use Permits	-	0.00	-	0.00
Traffic Control/Barricades	-	0.00	-	0.00
Dewatering	-	0.00	-	0.00
Test Equipment	-	0.00	-	0.00
Temporary Office Janitorial	-	0.00	-	0.00
Total Temporary Facilities & Support	-	0.00	-	0.00
TEMPORARY UTILITIES				
Install Temporary Power Service	-	0.00	-	0.00
Temporary Power Bills	-	0.00	-	0.00
Temporary Power Distribution	-	0.00	-	0.00
Temporary Lighting	-	0.00	-	0.00
Temporary Water Bills	-	0.00	-	0.00
Temporary Water Distribution	-	0.00	-	0.00
Portable Toilets	-	0.00	-	0.00
Temporary Heat	-	0.00	-	0.00
Total Temporary Utilities	-	0.00	-	0.00

CONTRACTOR C Total SF: 340,000		NOTES
$	$/SF	
-	0.00	
-	0.00	
-	0.00	
-	0.00	
-	0.00	
-	0.00	
-	0.00	
-	0.00	
-	0.00	
-	0.00	
-	0.00	
-	0.00	
-	0.00	
-	0.00	
-	0.00	
-	0.00	
-	0.00	
-	0.00	
-	0.00	
-	0.00	
-	0.00	
-	0.00	
-	0.00	
-	0.00	
-	0.00	
-	0.00	
-	0.00	
-	0.00	
-	0.00	
-	0.00	
-	0.00	
-	0.00	
-	0.00	
-	0.00	
-	0.00	
-	0.00	

FIGURE 4-2 (CONTINUED)

XYZ CORPORATION BUILDING	CONTRACTOR A Total SF: 340,000		CONTRACTOR B Total SF: 340,000	
Description	$	$/SF	$	$/SF
HOISTING/FORKLIFTS				
Mobile Crane Rental	-	0.00	-	0.00
Mobile Crane Operator	-	0.00	-	0.00
Forklift Rental	-	0.00	-	0.00
Forklift Operator	-	0.00	-	0.00
Fuel & Maintenance	-	0.00	-	0.00
Total Hoisting/Forklifts	-	0.00	-	0.00
CONSTRUCTION EQUIPMENT				
Construction Equipment (attach detail)	-	0.00	-	0.00
Fuel & Maintenance	-	0.00	-	0.00
Small Tools	-	0.00	-	0.00
Consumables	-	0.00	-	0.00
Shop Foreman	-	0.00	-	0.00
Freight & Trucking	-	0.00	-	0.00
Total Construction Equipment	-	0.00	-	0.00
PERSONNEL/MATERIAL HOISTING				
Protect Elevator Doors & Fronts	-	0.00	-	0.00
Protect Elevator Cabs	-	0.00	-	0.00
Man/Material Hoist	-	0.00	-	0.00
Hoist Operator	-	0.00	-	0.00
Elevator Maintenance	-	0.00	-	0.00
Elevator Operator	-	0.00	-	0.00
Total Personnel/Material Hoisting	-	0.00	-	0.00
CLEANUP				
Continuous Cleanup	-	0.00	-	0.00
Final Cleanup	-	0.00	-	0.00
Disposal Bills	-	0.00	-	0.00
Window Cleaning	-	0.00	-	0.00
Total Cleanup	-	0.00	-	0.00
PROJECT CLOSEOUT				
As-builts	-	0.00	-	0.00
Operation & Maintenance Manuals	-	0.00	-	0.00
Total Project Closeout	-	0.00	-	0.00
Totals:	**0.00**	**0.00**	**0.00**	**0.00**

Notes/Comments: _____

_____

_____

_____

CONTRACTOR C Total SF: 340,000		NOTES
$	**$/SF**	
-	0.00	
-	0.00	
-	0.00	
-	0.00	
-	0.00	
-	0.00	
-	0.00	
-	0.00	
-	0.00	
-	0.00	
-	0.00	
-	0.00	
-	0.00	
-	0.00	
-	0.00	
-	0.00	
-	0.00	
-	0.00	
-	0.00	
-	0.00	
-	0.00	
-	0.00	
-	0.00	
-	0.00	
-	0.00	
-	0.00	
-	0.00	
-	0.00	
0.00	**0.00**	

The AIA A111™ Agreement and associated AIA A201™, *General Conditions,* specifically define costs to be included in the work. The owner may choose to include additional cost elements, which should be identified in the RFP. Verify that the contractor's response is consistent within these parameters. This will reduce the potential for disagreements and cost surprises later in the negotiation process.

Section C: Construction Agreement

It is imperative to include in the RFP the construction contract in its entirety. The contract is the foundation of the deal; its terms and conditions may affect the contractor's fee, general requirements, markups, and other relevant proposed costs. Request a detailed description of all exceptions to the contract. This is crucial in eliminating misunderstandings and disagreements during the negotiation phase, which can become contentious and cause delays.

Section D: Pricing Package and Documents

1. PROJECT DOCUMENTS

Traditionally, the negotiated RFP is issued early in the design process. It can, however, be effectively issued anytime along the continuum of conceptual to design development documents. Therefore, the extent of the drawings available for distribution with the RFP can vary. At a minimum, basic floor plans, elevations, building section(s), and a site plan should be included.

Additional documents to be considered include

➤ Architectural drawings

➤ Room finish schedule

➤ Structural drawings

> Geotechnical report

> Environmental survey

> Mechanical, electrical, and plumbing drawings

> Site survey

2. NARRATIVE(S) AND/OR OUTLINE SPECIFICATIONS

It is important to develop an outline specification that includes a level of detail sufficient to underpin the pro forma and program requirements. It may be helpful to include one or more narratives from the architect, engineers, and owner, defining essential elements such as exterior closure materials, type of structure, program requirements (e.g., square feet, units), mechanical and electrical systems, finish schedule, and site development details (e.g., pool, special landscaping/features). Increasing the contractor's understanding of the project increases the likelihood of receiving meaningful feedback in the response. (Refer to the Project Narrative in the RFP example included in the Appendix, for an example of narrative details.)

The scope of work may be further defined by providing a list of what is *not* included. A good starting place in developing exclusions is the pro forma. Items such as permits, FF&E, and design-build components can be costly and are often overlooked in the pricing exercise. Don't forget the all-important consideration: Who carries what tax responsibility? For example, taxes on equipment rental and payroll may be the responsibility of the contractor, but sales tax on the overall project may be the responsibility of the owner. In the state of Washington, sales tax on a $10 million project is $880,000—a significant impact to the pro forma if it is assumed to be included in the contractor's price!

By example, other project elements that may be overlooked include

> Utility services to the site (off-site)

> Municipal requirements for improvements to adjacent streets
(paving, lighting)

➤ Testing and inspections

➤ Certain engineering costs (e.g., shoring, survey)

➤ Design-build engineering and permit costs

➤ Street and use permits

➤ Municipal permits and inspection

➤ Systems sometimes furnished and installed by owner's vendors (e.g., telecommunications, security, data)

3. FORMS

Include appropriate forms for submission of quantitative information in consistent format. By providing the format to be used for comparative analysis the process will be efficient and timely. In the event the RFP is estimate-based, the forms should include the Cost Estimate Analysis Format and the GRs Format. If the RFP is fee-based include the GRs Format.

Summary

On larger projects, the RFP package will likely be extensive. Contributing the necessary resources to assemble a complete package will pay dividends. The documents will serve as the basis for a quantitative and thorough analysis— and increase the likelihood of selecting the best candidate. The package will benefit the entire owner-designer-contractor team throughout the design and budgeting process, providing a road map for the design and final GMP.

On smaller projects, while the extent of the requested information may be reduced, it remains important to provide adequate detail in the RFP package. Meaningful and reliable contractor responses, be they for smaller or larger projects, are dependent on the quality of information provided in the RFP which, in turn, provides a well-defined basis for the construction contract.

When the RFP is ready for issuance, consideration should be given to conducting an introductory meeting with all contractor candidates. Issue the

RFP at least a week before the kickoff meeting, to allow the candidates time to familiarize themselves with the project and the RFP requirements, and to compile relevant questions for discussion at the introductory meeting.

The introductory meeting provides the opportunity to present essential elements of the project and ensures that all candidates hear the same message, thereby maintaining consistency in defining the project and the related expectations and outcomes. Any questions presented by the candidates may be answered during the meeting so all have consistent direction and clarification. In this manner the answers will be presented to all candidates at the same time-allowing follow-up discussion and clarification of the issue at hand and thereby ensuring a fair process. It is essential to document the information and directives provided in this meeting, and distribute to all (if applicable) as an addendum to the RFP.

The RFP solicits data from the contractors that are integral to the final "deal." The more specific the RFP documents, the more revealing the responses. Specifics lead to a tight deal.

There are many elements to consider in analyzing responses, selecting the best candidate, and finalizing a tight deal. All are essential to the ultimate success of the project and must be carefully evaluated. Quantitative and qualitative analysis of the responses is discussed in Chapter 5.

CHAPTER 5

The Analysis

When the RFP includes specific instructions defining the format for the responses and the use of the associated electronic spreadsheets, the analysis can be efficient and timely. Checklists and spreadsheets should be prepared in advance of the due date to facilitate expedient comparison of the response data. If each contractor's response follows the defined format, the information can be readily inserted into the prepared checklists. When the format and preparations are well coordinated, it is possible to complete the analysis in a matter of days.

The objective of the analysis is to refine the responses to a meaningful and concise summary of information relating to the selection criteria. This may include general requirements costs, fees, markups, and other tangible information specific to the project.

Estimate-based Plan (Preferred Method)

If the RFP includes a request for a detailed construction estimate, the costs can be entered into a spreadsheet designed for quantitative cost comparison between contractors. A sample spreadsheet analyzing the contractors' estimated construction costs, fees, and related markups is presented in Figure 5-1.

Analysis of the spreadsheet data can be revealing. For example, the pricing provided by the contractors under the heading "Structure" shows a comparatively limited variance. This may indicate that the contractors have a relatively consistent understanding of the structure and/or reasonably clear documents. On the other hand, pricing identified under the heading "Interior Construction" varies considerably, which may indicate inconsistent understanding of the finishes and/or incomplete documents.

In such a case it is recommended that each contractor be questioned to assess assumptions, inclusions, and exclusions related to the specific project component(s), as well as an understanding of the scope of work. For example, you may find that some contractors include specific elements under different headings, which may explain the differences. It is essential to the reliability of the analysis to understand these differences, whatever they may be, and make the necessary adjustments in the spreadsheet.

Subsequent distribution of additional clarifications and/or information may be required. Appropriate documentation of these clarifications should be provided to all respondents with a request to adjust pricing if necessary. This is a powerful process in refining the design intent as well as the reliability of the pricing.

Pricing in the estimate-based plan is preliminary and based on schematic documentation; therefore, selection of the contractor based on lowest price is not reliable. The estimate-based analysis provides a means of discovery at a level of detail well beyond that in the fee-based approach (which assesses only the cost issues of fees, markups, and general requirements and which is described in more detail in the next section). The detailed estimate-based analysis can reveal the level of the technical capablities of each contractor and stimulate focused discussion about specific elements of the project. The follow-up questioning provides *unparalleled access* to all contractor candidates' cost databases. This vast knowledge base is not available after one contractor has been chosen. For this reason this effort will be highly beneficial to the project. Contributing the appropriate effort and resources during this phase is crucial.

Use the estimate-based analysis and follow-up to gain knowledge about cost and design options relating to your project and to determine the contrators' estimating capabilities, constructability knowledge, value engineering expertise, and creative thinking skills. In addition, the average of all estimates is a good first test of the validity of the assumed construction cost.

To assess the competitiveness of the estimate-based proposals, however, utilize the analysis tools presented in the fee-based plan (Figure 5-2) to evaluate fees, markups, general requirements, and related add-ons. By using the estimate-based analysis tools *and* the fee-based tools *together*, comparisons and evaluations will be informative, revealing, and fair.

Fee-based Plan

In the fee-based plan, the analysis of the various markups and fees can be confusing and misleading. For example, a proposal indicating a low fee with higher insurance rates may be more costly than another with a higher fee. Unless the fees and markups are assessed as a composite package, the analysis may not present an accurate assessment.

Figure 5-2 is an example of an analysis spreadsheet used to assess the combined effect of all markups in a fee-based plan. In the example, to assess the combined effect of the contractors' proposed financial terms (fees, markups, and general requirements), an allowance for each has been removed from the target construction cost. In the event the RFP requests general requirements costs, an allowance for GRs would likewise be removed. A reasonable allowance amount can be determined through discussion with the architect and/or contractor community. (The exact amount of the allowances is only important to the fair analysis of the proposals, as it establishes a consistent basis for the comparison.) It is most important that these allowed amounts be applied consistently for all proposals to reach a baseline of bare construction costs (no fees, markups, GRs). The subtotal ("Estimated Bare Construction Cost") then represents a uniform amount that serves as the basis for measurement of the combined effect of the markups.

FIGURE 5-1 Cost Estimate Analysis

Total Project Gross SF	340,000	SF	
Net Rentable SF	185,000	SF	
Number of Units	200	EA	

CONTRACTOR PROPOSALS	CONTRACTOR A			CONTRACTOR B		
		% of			% of	
UniFormat BREAKDOWN	Total	Subtotal	$/GSF	Total	Subtotal	$/GSF
STRUCTURE	$5,133,502	21.41%	$15.10	$5,398,720	21.98%	$15.88
EXTERIOR CLOSURE	$2,902,828	12.10%	$8.54	$1,977,520	8.05%	$5.82
ROOFING	$602,845	2.51%	$1.77	$308,000	1.25%	$0.91
INTERIOR CONSTRUCTION	$6,091,614	25.40%	$17.92	$6,993,440	28.48%	$20.57
CONVEYING SYSTEMS	$513,118	2.14%	$1.51	$604,800	2.46%	$1.78
PLUMBING SYSTEMS	$2,088,221	8.71%	$6.14	$2,331,840	9.50%	$6.86
HVAC SYSTEMS	$1,212,378	5.06%	$3.57	$1,344,000	5.47%	$3.95
FIRE PROTECTION	$501,364	2.09%	$1.47	$588,000	2.39%	$1.73
ELECTRICAL SYSTEMS	$2,690,424	11.22%	$7.91	$2,514,960	10.24%	$7.40
SITE WORK/DEMOLITION	$1,109,326	4.63%	$3.26	$882,080	3.59%	$2.59
GENERAL REQUIREMENTS	$1,136,700	4.74%	$3.34	$1,615,000	6.58%	$4.75
SUBTOTAL	$23,982,320	100.00%	$70.54	$24,558,360	100.00%	$72.23
	% of Total Construction Costs			% of Total Construction Costs		
Contractor's Fee	$508,000	2.00%	$1.49	$793,000	3.00%	$2.33
State & Local Taxes	$205,000	0.81%	$0.60	$215,000	0.81%	$0.63
Insurance	$215,000	0.85%	$0.63	$198,000	0.75%	$0.58
Contractor's Contingency	$508,000	2.00%	$1.49	$660,000	2.50%	$1.94
Other	$0	0.00%	$0.00	$0	0.00%	$0.00
TOTAL CONSTRUCTION COSTS	$25,418,320		$74.76	$26,424,360		$77.72
Escalation	$0	0.00%	$0.00	$0	0.00%	$0.00
Bond	$126,000	0.50%	$0.37	$290,000	1.10%	$0.85
Preconstruction Amount	$0	0.00%	$0.00	$0	0.00%	$0.00
TOTAL	$25,544,320		$75.13	$26,714,360		$78.57

| CONTRACTOR C | | | AVERAGE | | | NOTES |
Total	% of Subtotal	$/GSF	Total	% of Subtotal	$/GSF	
$5,661,583	25.37%	$16.65	$5,397,935	22.85%	$15.88	
$2,612,997	11.71%	$7.69	$2,497,782	10.58%	$7.35	Follow up Contractor B scope
$386,408	1.73%	$1.14	$432,418	1.83%	$1.27	
$4,961,117	22.23%	$14.59	$6,015,390	25.47%	$17.69	Follow up Contractor C scope
$649,331	2.91%	$1.91	$589,083	2.49%	$1.73	
$1,732,430	7.76%	$5.10	$2,050,830	8.68%	$6.03	Follow up Contractor C scope
$962,480	4.31%	$2.83	$1,172,953	4.97%	$3.45	Follow up Contractor C scope
$419,475	1.88%	$1.23	$502,946	2.13%	$1.48	
$2,231,941	10.00%	$6.56	$2,479,108	10.50%	$7.29	
$1,539,880	6.90%	$4.53	$1,177,095	4.98%	$3.46	Follow up Contactor B scope
$1,156,374	5.18%	$3.40	$1,302,691	5.52%	$3.83	Follow-up required/see GRs analysis
$22,314,016	100.00%	$65.63	$23,618,232	100.00%	$69.47	
% of Total Construction Costs			% of Total Construction Costs			
$528,000	2.25%	$1.55	$609,667	2.43%	$1.79	
$223,000	0.95%	$0.66	$214,333	0.85%	$0.63	
$90,000	45.00%	$0.00	$167,667	0.67%	$0.00	Contractor C cost included above
$235,000	1.00%	$0.69	$467,667	1.86%	$1.38	
$35,000	0.15%	$0.10	$11,667	0.05%	$0.03	Contractor C: "Info. Technology Fee"
$23,425,016		$68.90	$25,089,232		$73.79	
$0	0.00%	$0.00	$0	0.00%	$0.00	Discuss with team
$200,000	0.85%	$0.59	$205,333	0.82%	$0.60	
$50,000	0.00%	$0.15	$16,667	0.07%	$0.05	Contractors A and B follow-up
$23,675,016		$69.63	$25,311,232		$74.44	

FIGURE 5-2 Fee and Markups Analysis

Total Project Gross SF	340,000	SF
Net Rentable SF	185,000	SF
Number of Units	200	EA

UPDATED FEE PROPOSALS	CONTRACTOR A			CONTRACTOR B		
	Amount	% of Pro Forma Construction Amount	$/SF	Amount	% of Pro Forma Construction Amount	$/SF
1 Pro Forma Construction Cost	$25,000,000	100.00%	$73.53	$25,000,000	100.00%	$73.53
2 Less Estimated Fee & Markups (6%)	–$1,500,000	–6.00%	–$4.41	–$1,500,000	–6.00%	–$4.41
3 Less Estimated General Requirements (6%)	–$1,500,000	–6.00%	–$4.41	–$1,500,000	–6.00%	–$4.41
Estimated Bare Construction Cost	$22,000,000	88.00%	$64.71	$22,000,000	88.00%	$64.71
	Quoted % or Lump Sum Amount	Quoted % or Lump Sum Amount		Quoted % or Lump Sum Amount	Quoted % or Lump Sum Amount	
4 General Requirements	$1,136,700	5.17%	$3.34	$1,615,000	7.34%	$4.75
SUBTOTAL	$23,136,700			$23,615,000		
5 Contractor's Fee	$490,496	2.00%	$1.44	$762,266	3.00%	$2.24
6 State & Local Taxes	$198,651	0.81%	$0.58	$205,812	0.81%	$0.61
7 Insurance	$208,461	0.85%	$0.61	$190,566	0.75%	$0.56
8 Contractor's Contingency	$490,496	2.00%	$1.44	$635,222	2.50%	$1.87
10 Other	$0	0.00%	$0.00	$0	0.00%	$0.00
TOTAL FEE & MARKUP COMPARISON	$24,524,804	100.00%	$72.13	$25,408,866	100.00%	$74.73
11 Escalation	$0	0.00%	$0.00	$0	0.00%	$0.00
9 Bond	$122,624	0.50%	$0.36	$279,498	1.10%	$0.82
12 Preconstruction Amount or % Fee	$0	0.00%	$0.00	$0	0.00%	$0.00
TOTAL	$24,647,428		$72.49	$25,688,363		$75.55

Notes:
Line 1:   Target construction cost from pro forma.
Lines 2 & 3:   Remove estimated markups and general requirements to establish baseline for comparison of proposed fee and markups.
Line 4:   Actual proposed general requirements from proposals.
Lines 5–10:   Enter quoted lump sum or % from contractors' proposals.
Line 11:   Owner, architect, and possibly contractor determine if escalation calculation is appropriate depending on market conditions.

for finishes protection. In the event the project includes features finishes, it may be advisable to question the contractor's approach to ensuring quality in the finished product.

Used together, the analysis formats presented in Figures 5-1, 5-2, and 5-3 provide informative comparisons of the responses. These not only are valuable in assessing the cost impact of the proposed financial elements, but also may reveal the extent of each contractor's knowledge about this type of project and project management style.

Follow-up

Follow-up with the contractors may be required to complete the analysis. You should not hesitate to contact each candidate to further understand the basis of the proposal and the elements of cost. Figure 5-4 is an example of a follow-up document.

After completion of the worksheets in Figures 5-1, 5-2, and 5-3, follow-up with the contractors is conducted. The follow-up conversations may reveal omissions, misunderstandings, scope clarifications, and related issues affecting the proposed costs. In the follow-up document presented in Figure 5-4, the clarifications are documented in writing and issued to all contractors for possible cost adjustment. Responses from the contractors are requested in writing to maintain the paper trail required at final negotiation.

Upon completion of the follow-up effort, final adjustments may be required to assess the impact of the new information on the final estimated construction cost. This exercise is the culmination of the financial analysis, utilizing the refined contractor input.

Figure 5-5 illustrates how responses to the follow-up inquiry may affect the cost analysis.

The construction cost is carried forward from the analysis forms presented in Figure 5-1 or 5-2. Note that the example includes follow-up categories related to an estimate-based RFP (Figure 5-1).

FIGURE 5-3 # General Requirements Analysis

XYZ CORPORATION BUILDING	CONTRACTOR A Total SF: 340,000		CONTRACTOR B Total SF: 340,000	
Description	$	$/SF	$	$/SF
PROJECT STAFF & SUPPORT				
Project Manager	125,000	0.37	135,000	0.40
Project Superintendent	125,000	0.37	125,000	0.37
Project Engineer	92,000	0.27	150,000	0.44
Cost/Scheduling Engineer	-	0.00	75,000	0.00
Office Administrator	30,000	0.09	36,000	0.11
Travel & Subsistence	-	0.00	2,000	0.01
Project Legal Fees	-	0.00	5,000	0.01
Professional Engineering Services	-	0.00	10,000	0.03
Scheduling	-	0.00	5,000	0.01
Total Project Staff & Support	372,000	1.09	543,000	1.38
PROJECT SAFETY/SECURITY				
Safety Training	-	0.00	5,000	0.01
Safety Engineer	15,000	0.04	20,000	0.06
Security/Watchman	-	0.00	32,000	0.09
Security Equipment	-	0.00	5,000	0.01
Total Project Safety/Security	15,000	0.04	62,000	0.18
FIELD SUPERVISION & FIELD ENGINEERING				
Field Supervision	145,000	0.43	175,000	0.51
Field Engineering	80,000	0.24	85,000	0.25
Engineering Equipment	16,000	0.05	12,000	0.04
Engineering Supplies	-	0.00	5,000	0.01
As-built Documents	2,000	0.01	4,000	0.01
Quality Control	-	0.00	20,000	0.06
Certified Survey	-	0.00	5,000	0.01
Total Field Supervision & Field Engineering	243,000	0.71	306,000	0.90
PROJECT OFFICE				
Office Equipment	1,200	0.00	1,500	0.00
Office Supplies	12,000	0.04	8,000	0.02
Telephone/Data	7,500	0.02	6,000	0.02
Postage & Shipping	500	0.00	2,000	0.01
Copies/Reproduction	1,000	0.00	2,000	0.01
Drinking Water	250	0.00	-	0.00
Project Office Setup/Dismantle	20,000	0.06	5,000	0.01

CONTRACTOR C Total SF: 340,000		NOTES
$	$/SF	
108,000	0.32	
98,000	0.29	
84,000	0.25	
-	0.00	
28,000	0.08	
-	0.00	
-	0.00	
5,000	0.01	
2,600	0.01	
325,600	0.96	
-	0.00	
12,000	0.04	
-	0.00	
2,250	0.01	
14,250	0.04	Contractors A and C follow-up
133,800	0.39	
76,800	0.23	
5,000	0.01	
8,500	0.03	
-	0.00	
9,550	0.03	
5,500	0.02	
239,150	0.70	
500	0.00	
11,000	0.03	
8,550	0.03	
1,200	0.00	
1,250	0.00	
-	0.00	
12,550	0.04	

FIGURE 5-3 (CONTINUED)

XYZ CORPORATION BUILDING	CONTRACTOR A Total SF: 340,000		CONTRACTOR B Total SF: 340,000	
Description	$	$/SF	$	$/SF
PROJECT OFFICE (cont.)				
Project Office Rental/Maintenance	-	0.00	3,000	0.01
Copy Machine	-	0.00	1,200	0.00
Computers	2,000	0.01	6,000	0.02
Printers & Fax Machine	-	0.00	1,700	0.01
Furniture & Equipment	-	0.00	2,500	0.01
Total Project Office	44,450	0.13	38,900	0.11
TEMPORARY FACILITIES & SUPPORT				
Trailer Rental	-	0.00	8,500	0.03
Temporary Fence & Maintenance	5,000	0.01	4,000	0.01
Shop Costs	2,000	0.01	1,200	0.00
Tool & Dry Sheds	-	0.00	5,000	0.01
Project Signs	3,250	0.01	300	0.00
Temporary Guardrails	20,000	0.06	24,000	0.07
Safety Equipment & Supplies	12,000	0.04	18,000	0.05
Weather Protection	5,000	0.01	8,000	0.02
Temporary Stair Towers	-	0.00	4,500	0.01
Temporary Ladders	-	0.00	5,000	0.01
Temporary Fire Protection	-	0.00	3,000	0.01
Protect Finishes	12,000	0.04	-	0.00
Temporary Roads/Parking	10,000	0.03	5,500	0.02
Street Sweeping/Dust Control	-	0.00	2,500	0.01
Street and Use Permits	-	0.00	2,000	0.01
Traffic Control/Barricades	10,000	0.03	12,000	0.04
Dewatering	3,000	0.01	5,000	0.01
Test Equipment	-	0.00	3,000	0.01
Temporary Office Janitorial	-	0.00	1,200	0.00
Total Temporary Facilities & Support	82,250	0.24	112,700	0.33
TEMPORARY UTILITIES				
Install Temporary Power Service	75,000	0.22	95,000	0.28
Temporary Power Bills	16,000	0.05	17,000	0.05
Temporary Power Distribution	-	0.00	16,000	0.05
Temporary Lighting	12,000	0.04	15,000	0.04
Temporary Water Bills	6,000	0.02	5,000	0.01
Temporary Water Distribution	1,200	0.00	-	0.00
Portable Toilets	6,000	0.02	3,500	0.01
Temporary Heat	20,000	0.06	16,000	0.05
Total Temporary Utilities	136,200	0.40	167,500	0.49

CONTRACTOR C Total SF: 340,000		NOTES
$	$/SF	
6,500	0.02	
-	0.00	
-	0.00	
-	0.00	
1,200	0.00	
42,750	0.13	
5,500	0.02	
8,300	0.02	
-	0.00	
3,200	0.01	
-	0.00	
18,500	0.05	
3,364	0.01	
7,200	0.02	
3,000	0.01	
2,000	0.01	
12,300	0.04	
7,080	0.02	Contractor B no finishes protected?
8,600	0.03	
3,200	0.01	
1,800	0.01	Contractor A not per RFP?
3,500	0.01	
1,200	0.00	
-	0.00	
-	0.00	
88,744	0.26	
80,200	0.24	
12,500	0.04	
12,500	0.04	
17,500	0.05	
2,500	0.01	
2,200	0.01	
3,200	0.01	
19,500	0.06	
150,100	0.44	

FIGURE 5-3 (CONTINUED)

XYZ CORPORATION BUILDING	CONTRACTOR A Total SF: 340,000		CONTRACTOR B Total SF: 340,000	
Description	$	$/SF	$	$/SF
HOISTING/FORKLIFTS				
Mobile Crane Rental	25,000	0.07	35,000	0.10
Mobile Crane Operator	14,000	0.04	10,200	0.03
Forklift Rental	13,600	0.04	12,000	0.04
Forklift Operator	6,000	0.02	4,500	0.01
Fuel & Maintenance	5,000	0.01	3,500	0.01
Total Hoisting/Forklifts	63,600	0.19	65,200	0.19
CONSTRUCTION EQUIPMENT				
Construction Equipment (attach detail)	1,000	0.00	58,000	0.17
Fuel & Maintenance	6,000	0.02	13,000	0.04
Small Tools	2,500	0.01	3,500	0.01
Consumables	3,000	0.01	2,800	0.01
Shop Foreman	6,500	0.02	-	0.00
Freight & Trucking	6,000	0.02	14,000	0.04
Total Construction Equipment	25,000	0.07	91,300	0.27
PERSONNEL/MATERIAL HOISTING				
Protect Elevator Doors & Fronts	-	0.00	2,500	0.01
Protect Elevator Cabs	-	0.00	1,200	0.00
Man/Material Hoist	24,000	0.07	45,000	0.13
Hoist Operator	12,000	0.04	16,000	0.05
Elevator Maintenance	2,500	0.01	1,200	0.00
Elevator Operator	8,000	0.02	12,500	0.04
Total Personnel/Material Hoisting	46,500	0.14	78,400	0.23
CLEANUP				
Continuous Cleanup	55,000	0.16	78,000	0.23
Final Cleanup	25,000	0.07	28,000	0.08
Disposal Bills	23,000	0.07	36,500	0.11
Window Cleaning	-	0.00	2,000	0.01
Total Cleanup	103,000	0.30	144,500	0.43
PROJECT CLOSEOUT				
As-builts	2,500	0.01	3,000	0.01
Operation & Maintenance Manuals	3,200	0.01	2,500	0.01
Total Project Closeout	5,700	0.02	5,500	0.02
Totals:	**1,136,700**	**3.34**	**1,615,000**	**4.75**

Notes/Comments: Follow-up required:

Contractor A: Street use permits? Safety and security?

Contractor B: Protection of finishes?

Contractor C: As-built and O&M documents? Safety and security?

CONTRACTOR C		NOTES
Total SF: 340,000		
$	$/SF	
41,000	0.12	
8,500	0.03	
5,000	0.01	
2,500	0.01	
2,000	0.01	
59,000	0.17	
19,380	0.06	
10,500	0.03	
2,500	0.01	
8,000	0.02	
2,800	0.01	
15,000	0.04	
58,180	0.17	
1,200	0.00	
500	0.00	
33,000	0.10	
9,500	0.03	
2,800	0.01	
4,800	0.01	
51,800	0.15	
83,600	0.25	
21,200	0.06	
22,000	0.06	
-	0.00	
126,800	0.37	
-	0.00	
-	0.00	
-	0.00	Contractor C not providing as-builts/O&M's?
1,156,374	**3.40**	

FIGURE 5-4 **Follow-up Document**

The Fria Company, Inc.

e-mail: rick@friaCM.com

(Addressee)

We have reviewed proposals and pricing for the XYZ Corporation multifamily project and require further information in order to complete a fair and meaningful assessment. In order to refine the proposals, please adjust your pricing to reflect the following scope clarifications:

1. The unit mix has been revised and reduced from 205 to 200. A new unit mix is provided with this memo. Please note there are less upgraded units and twelve (instead of ten) penthouse units. Revised drawings are being sent to you this week that reflect these revisions. These changes do not reduce the net rentable (finished) SF but do reduce kitchens, bathrooms, plumbing fixtures (plumbing fixture schedule not revised), doors, etc. Please revise the finishes, mechanical, and other relevant portions of your estimate accordingly.

2. Exterior wall is painted (assume two-coat elastomeric).

3. Windows are either vinyl or coated aluminum, whichever represents your best pricing. Be sure to consider the energy performance characteristics of the windows required by the use of electric resistant heating in the floors below the penthouse units. The Portland Energy Code will dictate upgraded performance of the glass.

4. Electrical systems should include individual metering of each apartment and the electric resistance heating. A $500-per-unit lighting upgrade allowance should be included for the upgraded and penthouse units. Also included should be voice-entry access control (reporting from the main street entrance to each apartment), card-access security in the garage and at the main public entries, fire and life safety. Cable TV and telephone wiring should be excluded, but conduit should be included.

5. Plumbing should reflect the fixtures indicated in the finish schedule (not revised). Central gas-fired hot water should be included with meter slots (unions) in each apartment for individual metering of both hot and cold water.

6. Please separate mechanical costs to reflect fire sprinklers, HVAC, and plumbing.

7. For shoring, assume a $32/SF system, over 22,700 SF. This may not be reflective of the documents, but we have enough information relative to shoring to be comfortable with this number.

8. The shoring number includes water "depressurization" system (installed by the shoring contractor) intended to control the water entering the shored wall and, therefore, the excavation. In order to keep the numbers consistent, we ask you to include dewatering costs of $75,000 as an allowance.

We have not shared your pricing with anyone outside the ownership group. Please acknowledge in writing that your new budget reflects your current understanding of this scope.

Submit pricing revisions to The Fria Company by April 16.

In the example, follow-up findings affecting pricing include these:

➤ The cost for "Interior Construction" (see Figure 5-1) submitted by contractor C is notably low when compared to the others. Follow-up revealed contractor C had omitted the cost of the penthouse finishes, assuming the units were unfinished "shelled" space. The associated cost adjustment returned by contractor C is entered in the Final Proposal Analysis worksheet presented in Figure 5-5.

➤ The cost for "Plumbing Systems" submitted by contractor C is notably low compared to the others. Follow-up revealed the cost was based on hot water heaters in each apartment rather than the intended central hot water heating system. The associated cost adjustment returned by contractor C is again entered in the Final Proposal Analysis worksheet.

➤ Follow-up investigation found that under "Preconstruction," contractor C included the cost of preconstruction in the estimate, and contractors A and B did not. To assess the comparable costs, adjustments are made on this sheet, which affect the totals.

In a fee-based RFP plan, the amount carried forward from the form in Figure 5-2 is entered on the same line, but the follow-up categories vary from the example. Examples of follow-up categories for a fee-based RFP include

➤ General requirements adjustments related to review of the GRs spreadsheet

➤ Preconstruction fee adjustments related to follow-up inquiry

➤ Escalation adjustments based on team discussions

➤ Other cost adjustments resulting from the analysis and follow-up effort

Remember the bottom-line total in the fee-based approach is useful in assessing the financial elements of the proposals, but is not a reliable measure of cost since estimates were not submitted.

FIGURE 5-5 Final Proposal Analysis

ITEM/DESCRIPTION		CONTRACTOR A			CONTRACTOR B		
		UNITS	INCLUDED?	ADD'L COST	UNITS	INCLUDED?	ADD'L COST
TOTAL ESTIMATE AMOUNT				$25,544,320			$26,714,360
Adjustments							
Schedule Adjustment: 14 Month @ $/Month	$125,000	14		$0	16		$250,000
Safety/Security			YES	$20,000		YES	$0
Escalation		1.00%	NO	$255,443	1.00%	NO	$267,144
Preconstruction Cost		Lump Sum	NO	$100,000	Lump Sum	NO	$90,000
Mechanical/Electrical Design-Build Engineering Costs			YES	$0	Lump Sum	NO	$65,000
Street Use Permits			NO	$4,500		YES	$0
Plumbing Central Hot Water			YES	$0		YES	$0
Penthouses Interior Finishes			YES	$0		YES	$0
Window Washing Equipment			YES	$0	Lump Sum	NO	$100,000
Protect Finishes			YES	$0	Lump Sum	NO	$25,000
Signage, Directories & Displays			NO	$70,000		NO	$70,000
Off-site Improvements (traffic lights, paving)			NO	$0		NO	$0
Adjustments Total				$449,943			$867,144
ADJUSTED CONSTRUCTION BUDGET				$25,994,263			$27,581,504
Sales Tax?	0.00%			$0			$0
Total Adjusted Construction Budget				$25,994,263			$27,581,504

	Contractor A	Contractor B	Contractor C
Score from Response Content Analysis	71	75	68
EVALUATION SUMMARY (1=worst/5=best)			
Conceptual Budget: Rank	4	1	5
Construction Fee: Rank	5	1	4
Experience in Product Type	3	4	5
Financial Qualifications	5	3	3
Construction Manager's Rating (after Interview)	0	0	0
Engineer's Rating (after Interview)	0	0	0
Architect's Rating (after Interview)	0	0	0
Owner's Rating (after Interview)	0	0	0
TOTALS	**88 Points**	**84 Points**	**85 Points**

CONTRACTOR C			NOTES
UNITS	INCLUDED?	ADD'L COST	
		$23,675,016	From Cost Estimate Analysis
			(or from Fee Estimate Analysis)
19		$625,000	Financing Cost: $125,000/mo.
	YES	$20,000	Add for watchman per owner
1.00%	NO	$236,750	Owner adjustment at 1%
	YES	$0	
Lump Sum	NO	$20,000	
	YES	$0	From Contractor A follow-up
	NO	$225,000	Contractor C had water tanks in apts.
Lump Sum	NO	$730,000	
Lump Sum	NO	$100,000	
	YES	$0	Contractor B follow-up add
	NO	$70,000	Allowance assigned by owner
	YES	−$55,000	In owner budget
		$1,971,750	
		$25,646,766	Includes schedule impact on financing
		$0	No sales tax
		$25,646,766	Includes schedule impact on financing

NOTES:

Gross Square Feet Area	**340,000**
Net Rentable Area	**185,000**
Number of Units	**200**

	Contractor A	**Contractor B**	**Contractor C**
Cost per Gross Square Feet w/out Sales Tax	$76.45	$81.12	$75.43
Cost per Net Rentable Area w/out Sales Tax	$140.51	$149.09	$138.63
Cost per Unit w/out Sales Tax	$129,971	$137,908	$128,234

Under either the fee-based or estimate-based approach, these final adjustments allow refinement of the proposed financial terms, which provides a basis for identifying competitive proposals and the final negotiation.

Cost of Proposed Schedule

The analysis form in Figure 5-5 is also useful in assessing the financial impact of the contractors' proposed schedules. On the line entitled "Schedule Adjustments" contractors B and C have proposed longer schedules than assumed in the pro forma, which will add financing interest cost for the additional time. In the "Notes" column, the monthly "Financing Cost" (interest) is indicated at $125,000. Since the pro forma assumes a 14-month schedule, there is no adjustment for contractor A; however, contractors B and C have been adjusted to account for the longer proposed schedules. This one adjustment can significantly impact the totals and should not be overlooked in the final cost analysis. The proposed schedule may be subject to negotiation later in the process, which is discussed in Chapter 7.

Response Content Analysis

Content and format of the responses can be revealing and may likewise be evaluated in a checklist. Consider these examples:

➤ Does the contractor follow directions? Was the proposal submitted on time? Was all the requested information included? Were the proper formats utilized and carefully followed?

➤ Is the response organized and complete? Is the proposal organized exactly as prescribed in the RFP? Does the proposal contain all the requested information?

➤ Is it presented in a professional manner? Is the package neatly bound or loose? Are there grammatical and punctuation errors in the documents? Is the information formatted and presented electronically (indicating the sophistication of the contractor's technological expertise)?

➤ Does the contractor indicate commitment? Is there a clear sense that the contractor is focused on your project? Is there a strong indication of desire to be a team player?

➤ Has the contractor made an effort to understand the owner's needs and goals? Did the contractor make an effort to determine the primary features of the project (e.g., highly finished entry lobby or special landscaping features) that will affect the success of the project's intended use or leasing?

The answers to these types of questions can be revealing. The checklist in Figure 5-6 is designed to assess these content and format issues. The spreadsheet should be modified to reflect the specific nuances of your project. The findings can also be useful during the interviews in composing questions for the finalists.

In the example, points are allotted for each RFP category to indicate the contractor's comparative strengths, capabilities, and expertise. The total points are carried forward to the Final Proposal Analysis (Figure 5-5) and entered under "Evaluation Summary" found on the bottom half of the sheet. The remaining categories are assigned points based on adjusted costs (ranking) and other critical elements of the plan. Final points are added by owner, architect, and other team members based on the interview outcome (discussed in Chapter 6).

Selecting Finalists

As the responses and spreadsheets are analyzed, it will become increasingly evident which contractors are most qualified and preferred. Once the information is compiled, it can be reviewed with the architect, engineers, and other personnel as desired. The goal is *to build, by consensus, a team of personnel who work well together.* By including the existing members of your team in the contractor evaluation process, the all-important team buy-in is more likely. The objective of the team review is to identify a "short list" for the interview process. This may be difficult, and in some cases the team may choose to interview all candidates.

FIGURE 5-6 # RFP Response Content Analysis

To complete this form, insert number ratings from 1 to 5 to grade the quality
of the response and/or the qualifications of the contractor.

Qualifications	CONTRACTOR A Included? YES NO POINTS	CONTRACTOR B Included? YES NO POINTS	CONTRACTOR C Included? YES NO POINTS	Notes
> Preconstruction services	x 0	x 0	x 3	Full services or limited?
> Letter of commitment	x 3	x 5	x 5	Company letterhead? All three parts included?
> Staffing/resumes	x 5	x 3	x 2	Quantity and experience
> References	x 3	x 5	x 5	Complete list? Positive feedback?
> Similar recent projects	x 3	x 4	x 5	Extent of experience in product type?
> Work to be self-performed	x 0	x 5	x 2	Beneficial to project? All identified?
> Proposed fee	x 4	x 1	x 5	Ranking
> Proposed general requirements costs	x 5	x 1	x 4	Ranking
> Schedule	x 5	x 3	x 1	Ranking
> Breakdown of pricing & format	x 5	x 4	x 2	Quality of product and complete?
> Qualifications, assumptions, and exclusions	x 5	x 5	x 5	Included? Red flags?
> Insurance certificate	x 3	x 5	x 4	Included? Limits OK?
> Bonding letter	x 4	x 5	x 3	Included? Surety letterhead? Quality of surety?
> Technical abilities	x 5	x 5	x 3	Demonstrated abilities?
> QC program	x 0	x 3	x 5	Included? Quality of program?
> Safety standards/performance	x 4	x 5	x 4	Does documentation indicate safety commitment?
> Team skills	x 3	x 3	x 5	Quality of skills for key members
> Cost control systems	x 4	x 5	x 0	Examples included? Effective system?
> Site visit: Examine job site	x 5	x 0	x 0	Did they make a site visit?
General Assessment				
> Were all requests in RFP addressed?	x 2	x 3	x 3	Complete and organized?
> Is proposal specific to this project?	x 3	x 5	x 2	Specific or generic material included in response?
Score for Response Content Analysis	71	75	68	**Carry forward to Final Proposal Analysis**

The examples included in this chapter were designed to assess RFP responses for an actual project and as such are formatted to include elements specific to that project. Your spreadsheets should be designed to accommodate the specific requirements of the project. Be sure to allow room for adjustments and flexibility as the RFP plan unfolds.

CHAPTER 6

The Interview

The interview will likely be the first time that you meet and interact with the contractor's proposed team. It is an essential component in the selection process therefore careful preparation is required to ensure maximum benefit.

Determine which consultants are of primary importance in the interview and selection process, and schedule their attendance. By including key members of your design and development team in the interview process, you will have an opportunity to assess the group dynamics and to promote team buy-in for the final selection.

Review the analysis results with the participants scheduled to attend the interview (architect, construction manager, relevant consultants) to ensure they are fully informed of the findings. This includes discussion of the concerns and issues requiring clarification by each contractor. The purpose of the preparatory review is to refine the interview plan to address specific issues relevant to each candidate. Each contractor RFP response is likely to present differing concerns and/or issues requiring resolution/clarification prior to final selection. By fine-tuning each interview you will be able to "fill in the blanks" in the short time allotted for this meeting.

If, following the RFP analysis, you have narrowed your choice to a short list of candidates, be sure (as a courtesy) to notify the contractors who have been eliminated. Plan to conduct all the interviews on the same day or on successive days, if possible. This will ensure a common thread through all interviews in comparing the contractors' proposed teams. Furthermore, the team discussion/decision can be made while the interview experiences are fresh and while all participants are focused on the decision process at hand.

A written notice of the interview date and time should be prepared and issued in advance.

Figure 6-1 is an example of such a notice. Be sure to allow adequate time for the contractors to prepare. The notice should include detailed interview requirements:

➤ Instruct the contractor to limit the number of attendees. In general, the fewer people, the more productive, meaningful, and interactive the interview. Suggest that a senior company representative be included to demonstrate contractor commitment. The proposed project manager, superintendent, and lead estimator should attend. There may be others whose attendance would add value to the interview process. Consult with your team to determine the appropriate mix for your project and objectives.

➤ Specify what you want to know. This could include details about schedule, budget, cost management, project reporting, value engineering, and subcontractor management/relations. For example, Figure 6-1 indicates key subjects and allotted time slots to be addressed at the interview. Note that it also allows time for an open forum during which the contractor may present relevant issues of her or his choosing, which may offer a revealing look at the firm and personnel.

FIGURE 6-1

The Fria Company, Inc.

e-mail: rick@friaCM.com

(Addressee)

Congratulations! Your firm has been selected as one of three finalists to construct the XYZ project. Please prepare for an interview with the team to discuss the issues outlined in this memo. Team members at the interview will include

T. Smith   Development Manager, XYZ Company
C. Williams Project Engineer, ABC Structural Engineering Company
D. Anderson   Principal, MMM Architecture Design
R. Fria   Construction Manager, The Fria Company

We will be conducting presentations/interviews on **May 4, in The Fria Company office**, as follows:

Contractor A   8:30 to 10 a.m.
Contractor B   10:30 to noon
Contractor C   1:30 to 3 p.m.

You will be held strictly to the 90-minute limit. Be prepared to present the following:

➤ **15 minutes:** Value engineering—how you do it, at what stages during the design you propose to do it, specific options for this project, and estimated values. This is a critical element of the presentation and selection.

➤ **15 minutes:** Scheduling, plan for building the project, and quality control discussion, all by the superintendent.

➤ **15 minutes:** Cost management, project management, and risk management methods, all by the project manager.

➤ **5 minutes:** Your plan for collaborative electronic networking (Web or otherwise) with the owner, design team, and owner's representative.

➤ **10 minutes:** Open forum for why your firm is the best selection.

➤ **20 minutes:** Open discussion and Q&A by owner representatives and you.

➤ **10-minute** grace allotted.

Your group should be limited to the project manager, superintendent, and estimator assigned to this project. A senior company representative should also attend.

Props are encouraged. Whiteboard, tackboard, and overhead projector are available. A model of the building and drawings will be in the room.

Creativity and thinking outside the box are encouraged.

- ➤ Specify the presentation tools and materials available in the interview room (e.g., overhead projector, whiteboards or tackboards, plans, building model). By informing the contractors in advance, you ensure that their preparations and presentations can be well orchestrated for maximum benefit. When the contractor is made aware of the issues to be addressed *and* the tools available to present the answers, you increase the chances of an effective and informative interview process.

- ➤ Identify a time slot for each contractor. Break down the time into planned increments, for example, introductions, 2 minutes; company history, 5 minutes; project specifics, 30 minutes; questions and answers, 15 minutes; grace period, 8 minutes.

- ➤ Allow adequate time between each interview for your team to discuss and rate the contractor's performance and to prepare for the next interview.

It is helpful to prepare questions for your team's use ahead of time. Figure 6-2 presents sample questions for use by the interview team. It is best to create the questions as a collaborative effort with the interview team, and to distribute these questions to the team in advance. The questions should be specific to the open issues requiring resolution by each contractor. Include a few questions that require the respondents to think on their feet; for instance, What do you think are the three most important elements of the project? You may even want to consider a "pop quiz" presenting a hypothetical problem or project challenge for the contractor team to resolve. This is a terrific opportunity to observe the interaction of the contractor's personnel, and it can reveal a great deal about the team to whom you may be entrusting the project. It can also reveal the extent of expertise of each team member. Often a right or wrong answer is less important than the process observed.

Meet with your team immediately prior to the interview to review potential questions and the desired selection criteria. This will help keep the issues fresh in team members' minds, thereby promoting a focused and meaningful interview process. It can also be beneficial to issue certain questions to the contractors in advance so that they can adequately prepare.

FIGURE 6-2 **Contractor Interview Questions**

XYZ CORPORATION PROJECT

General Questions:

➤ Will your project manager stay with the job from preconstruction through construction?
➤ Briefly describe your safety plan and your state safety rating.
➤ What is your experience working with mechanical design-build subcontractors?
➤ Clarify your intent of the contractor's contingency.

Questions for Project Manager:

➤ Briefly describe your experience with this product type.
➤ What are the three most important elements of the project as key to the success?
➤ Describe your plan for assembly of the GMP and buyout.
➤ Describe your budget-tracking process, and budget reporting to owner.
➤ Describe your buyout effort to ensure exterior wall quality control.
➤ How will you work with our team to ensure that your estimated costs can be maintained?

Questions for Superintendent:

➤ Briefly describe your experience with this product type.
➤ What are the three most important elements of the project as key to the success?
➤ Describe your quality control program for the exterior wall.
➤ Discuss the specific challenges of this site.

Specific Contractor A Questions:

➤ What are the specific project challenges, given the current local market?
➤ What is your backlog by project and volume for the next 2 years?
➤ For which major projects are you currently competing?
➤ Your GRs had very little cost allotted for safety and security. Please explain.

Specific Contractor B Questions:

➤ How will you perform preconstruction without a local office?
➤ Briefly define your plumbing and electrical estimates.
  Are they based on subcontractor input?
➤ Why is your insurance markup so low? Can you substantiate?
➤ Your GRs had no cost allotted for finishes protection.
  Please describe your approach to ensuring a quality final product.

Specific Contractor C Questions:

➤ What trades do you anticipate as open shop? Is there any conflict from unions?
➤ What is largest multifamily project your firm has constructed?
➤ What is your project manager's experience with design-build subcontractors?
➤ Your GRs had no cost allotted for operation and maintenance manuals and
  as-built documents. Please describe what you intend to provide at completion.

FIGURE 6-3 Final Proposal Analysis

ITEM/DESCRIPTION		CONTRACTOR A UNITS	INCLUDED?	ADD'L COST	CONTRACTOR B UNITS	INCLUDED?	ADD'L COST
TOTAL ESTIMATE AMOUNT				$25,544,320			$26,714,360
Adjustments							
Schedule Adjustment: 14 Month @ $/Month	$125,000	14		$0	16		$250,000
Safety/Security			YES	$20,000		YES	$0
Escalation		1.00%	NO	$255,443	1.00%	NO	$267,144
Preconstruction Cost		Lump Sum	NO	$100,000	Lump Sum	NO	$90,000
Mechanical/Electrical Design-Build Engineering Costs			YES	$0	Lump Sum	NO	$65,000
Street Use Permits			NO	$4,500		YES	$0
Plumbing Central Hot Water			YES	$0		YES	$0
Penthouses Interior Finishes			YES	$0		YES	$0
Window Washing Equipment			YES	$0	Lump Sum	NO	$100,000
Protect Finishes			YES	$0	Lump Sum	NO	$25,000
Signage, Directories & Displays			NO	$70,000		NO	$70,000
Off-site Improvements (traffic lights, paving)			NO	$0		NO	$0
Adjustments Total				$449,943			$867,144
ADJUSTED CONSTRUCTION BUDGET				$25,994,263			$27,581,504
Sales Tax?	0.00%			$0			$0
Total Adjusted Construction Budget				$25,994,263			$27,581,504

	Contractor A	Contractor B	Contractor C	Notes
Score from Response Content Analysis	71	75	68	See Figure 5-6
EVALUATION SUMMARY (1=worst/5=best)				
Conceptual Budget: Rank	4	1	5	
Construction Fee: Rank	5	1	4	
Experience in Product Type	3	4	5	
Financial Qualifications	5	3	3	
Construction Manager's Rating (after Interview)	5	1	0	C budgeting expertise poor
Engineer's Rating (after Interview)	3	4	0	B excellent structural team
Architect's Rating (after Interview)	5	0	3	A excellent staff
Owner's Rating (after Interview)	5	0	2	A best all around
TOTALS	**106 Points**	**89 Points**	**90 Points**	

CONTRACTOR C			NOTES
UNITS	INCLUDED?	ADD'L COST	
		$23,675,016	From Cost Estimate Analysis
			(or from Fee Estimate Analysis)
19		$625,000	Financing Cost: $125,000/mo.
	YES	$20,000	Add for watchman per owner
1.00%	NO	$236,750	Owner adjustment at 1%
	YES	$0	
Lump Sum	NO	$20,000	
	YES	$0	From Contractor A follow-up
	NO	$225,000	Contractor C had water tanks in apts.
Lump Sum	NO	$730,000	
Lump Sum	NO	$100,000	
	YES	$0	Contractor B follow-up add
	NO	$70,000	Allowance assigned by owner
	YES	–$55,000	In owner budget
		$1,971,750	
		$25,646,766	Includes schedule impact on financing
		$0	No sales tax
		$25,646,766	Includes schedule impact on financing

NOTES:			
Gross Square Feet Area	340,000		
Net Rentable Area	185,000		
Number of Units	200		
	Contractor A	**Contractor B**	**Contractor C**
Cost per Gross Square Feet w/out Sales Tax	$76.45	$81.12	$75.43
Cost per Net Rentable Area w/out Sales Tax	$140.51	$149.09	$138.63
Cost per Unit w/out Sales Tax	$129,971	$137,908	$128,234

When conducting the interview, strictly adhere to the planned schedule. At the beginning of each interview, outline what you expect and the time allowed. During the interview, it is important to let the contractor's personnel talk. Listen and watch for tangibles and intangibles, such as these:

➤ How does the team interact—with one another as well as with your team?

➤ Are they prepared?

➤ Are they professional in dress, manner, and stature?

➤ Do they know your project?

➤ Do they talk in generalities or specifics?

➤ Are they creative in offering solutions to specific project challenges?

➤ Are they enthusiastic about the project?

Immediately after each interview, while impressions are fresh, the interview team should devote adequate time to discussing and rating the various elements of the presentation. After all interviews are completed, numbers can be assigned by each interviewer, ranking overall preference or weighting the assessment of each contractor. Figure 6-3 presents an example of a ratings matrix for use during the interview process. In the example, many of the Final Proposal Analysis (FPA) entries originate during the RFP analysis phase (Chapter 5). This form is particularly useful during the interview process and should be distributed to each member of the team prior to interviews. It presents a summary-level overview of the critical analysis findings and may stimulate discussion during and after each interview.

When the points are totaled, the results will provide perspective to assist in *eliminating* candidates, but points should not be used as the sole basis for selection. Discussion following the interviews should include total points standings, fee and markups, qualifications, team dynamics, and other elements that represent the underlying goals of the project.

The people and the team dynamics will have considerable influence on the success of the project—and are as important as the fee. Cost alone is many times not the driving factor in contractor selection, especially in the event costs or financial elements are reasonably comparable.

Trust your instincts and intuition. How is the fit with the other team members? Consider the chemistry among the team. Are they team players? Do they communicate courteously and professionally? You will be working with this group for an extended time and entrusting them with your project, making the selection of a proactive, can-do team essential to the success of the project.

You will have the opportunity to negotiate terms (e.g., fee, markups, schedule) after completing the interviews. For example, if you like a specific contractor but the proposed fee is high, you can suggest a lower fee or creative alternatives as part of the follow-up and final negotiation.

Your team's input from the interview process, together with the RFP analysis, will provide the basis for the negotiation (Chapter 7) and the deal (Chapter 8).

The Negotiation

T he negotiation is the final chance to make your best deal. Remember, the RFP solicits "proposals" from the contractor candidates, intended to serve as the basis for the contract terms and the deal. As such, the proposed terms are subject to negotiation and refinement. Only after collecting and analyzing the data, interviewing the candidates, and reaching consensus with your team are you adequately prepared to negotiate the final terms.

At this point in the RFP plan, detailed information will have been gathered and analyzed. Face-to-face interviews will have afforded an opportunity to assess the intangibles. Presumably, the selection team will have reached consensus agreement on the most qualified and desired candidate(s). Although there may be a clear frontrunner, this is not always the case, and the negotiation may be conducted with more than one contractor. Either way, the negotiation should *appear* to include more than one contractor in order to maintain the highest degree of competitive response. Be sure your team agrees on the need for confidentiality, and to refrain from discussing the issues with outside parties, thereby creating the potential for compromised negotiations.

Analysis of the written proposals will have provided specific deal points such as fee, markups, labor rates (and burden), general requirements, and insurance rates. These are some of the key financial elements to use as a starting point for the negotiation. The interviews may have uncovered additional deal points. For example, the team may rate the project manager and the contractor highly, but may not have a positive impression of the superintendent. The negotiation presents an opportunity to request alternatives.

The team can identify points for negotiation, such as a lower percentage fee or a lump-sum fee, fixed general requirements in lieu of percentage basis, and other refinements to improve the deal.

As a general rule, ask for more than you hope to get. The team should determine the deal point "goal" prior to initiating the negotiation. This will allow the process to move forward in an organized fashion, delivering measurable results.

Be creative in formulating the negotiation, and maintain a spirit of give and take. Consider opportunities to "give" something to the contractor in exchange for "taking" a better deal. For example, a shorter schedule may be critical to the success of the project. To motivate the contractor to accept the shorter schedule, you may want to offer a higher fee or completion bonus. It is a simple cost/benefit analysis to assess the incremental increase of a higher fee against the cost-of-carry savings for a shorter schedule.

As the negotiation matures, be open to deal alternatives that may be unconventional yet provide win-win opportunities and support the essential project goals. Consultation with the selection team may reveal successful creative experiences on other projects that may not have been considered. It is critically important to apply the necessary time and effort to reaching an agreement, since this deal will serve as the basis for cost *and* the recipe for success.

It may be necessary to revise the analysis spreadsheets in order to accurately assess, in a quantitative manner, the affect of the proposed changes.

FIGURE 7-1

The Fria Company, Inc.

e-mail: rick@friaCM.com

(Addressee)

Dear Mr. Smith:

Thank you for a well-executed presentation Friday. It was a lively and interactive session, which answered many of our questions and concerns. We all enjoyed meeting your proposed team. Listening to others' views of the project is refreshing and revealing.

We are following up with three general contractor candidates for the XYZ Corporation multifamily project in Portland. We want to gather the last bits of information we feel are necessary to select the most qualified candidate and deliver the best value to XYZ Corporation. Please respond to the following issues today, if possible, and return a brief written response to my attention. You may respond directly on this form, if desired.

The initial issues are generic to all candidates. The last few are issues specific to your firm.

1. Even though your schematic pricing is inexact, will you accept a lump-sum fee fixed at the value indicated in your proposal? This means if the GMP is higher, your fee will remain the same provided scope remains the same.

2. Will you guarantee general requirements at the amount indicated in your proposal? Please attach a detailed summary of general requirements.

3. Please indicate your labor burden rate. Will you agree to fix the rate? Please forward the breakdown of inclusions (FUTA, SUTA, vacation, union fees, etc.).

4. Will you accept bond reimbursement at substantiated cost and at a rate not to exceed that indicated in your proposal?

5. Will you accept the insurance rate fixed at the rate indicated in your proposal?

6. Will you accept payment of the preconstruction cost indicated in your proposal at the time of construction start?

7. Should the project not proceed, would you accept payment of the preconstruction cost as a maximum amount, with less if substantiated as less?

8. Do you accept the reimbursable versus nonreimbursable cost description provided in the XYZ Corporation construction contract included in the RFP?

9. Will you agree that all home office expenses, including project-specific accounting, are nonreimbursable expenses? Will you agree that project-specific accounting will be done in the home office? Will you agree that time spent by the project executives (anyone other than the PM and superintendent) at the home office is not reimbursable?

10. Please provide documentation of your safety rating.

FIGURE 7-1 (CONTINUED)

11. Will you agree to competitively bid self-performed work, and if you win, will you handle as a subcontract amount if owner so chooses?

12. Will you accept Brown Drilling as the design/build shoring contractor, subcontracted directly to your firm?

13. Do you accept change-order markup limited to the amount of fee (%) indicated in your proposal? Subcontractor markups on COs limited to 15% OH&P combined?

14. Do you agree to provide the following interim budgets in detail?

    By June 15, a detailed budget update from your current budget, based on information gathered at the project kickoff meeting and the design review meetings.

    By September 15, a GMP based on 60% documents (design development).

    By December 15, a final GMP based on 100% documents (construction documents).

    Continual budget updates as required to account for preconstruction design decisions.

Contractor-specific issues, Contractor A:

A-1. We all liked John, your superintendent, but are concerned that he may not have the finishes expertise the project requires. Can you propose an assistant to John who has significant finishes expertise and can oversee the finishes process? Or, can you demonstrate that John has this level of expertise?

A-2. Will you accept your contractor's contingency at 1.5% in lieu of the 2.5% indicated in your proposal?

A-3. Please state your savings participation proposal.

A-4. Will you commit your staff as follows?

    Bob and John: Preconstruction as needed and full-time onsite during construction.

    Bruce: What percent of his time during preconstruction?

    Sam (MEP): As needed during preconstruction?

    Al: Full-time during preconstruction? During construction?

A-5. Will you agree to attend weekly design review/preconstruction coordination meetings in Portland? Will you agree to dedicate up to 2 full days for your preconstruction leader to be in Portland during preconstruction, if required by owner?

NOTE: We have scheduled May 25 for an all-day project review at the XYZ Corporation offices in Portland. Please pencil in this date for your project team, including superintendent, PM, project leader (executive), and estimator. In the event your team is selected, we want to get you up to speed immediately.

This may require review of the pro forma in order to consider "ripple" effects. For instance, the contractor may agree to perform preconstruction services without cost in order to win the project. If there is a preconstruction cost included in the pro forma, removing it may affect the bottom line. It is possible such an analysis may indicate that the contractor with a higher fee may present the best overall deal.

The most successful negotiations result in a win-win outcome for all parties. Beginning the project with a contractor who feels shortchanged in the negotiation may set the tone for a contentious relationship. The best outcomes for negotiated projects are a result of mutual respect and trust.

All communications during the negotiation should be documented. Figure 7-1 is an example of follow-up negotiation documentation. Several rounds of negotiation may be required and each should be memorialized and issued to the participant. In this manner each ensuing negotiation may proceed based on documented results from the previous discussion, and the potential for misunderstanding and confusion is greatly reduced. The goal is to end with an agreement that is well documented, includes clearly defined terms and conditions, and is viewed as a win-win for all. Documenting and finalizing the deal is discussed in the following chapter.

CHAPTER 8

The Deal

The culmination of the RFP effort is "the deal." By the time the team is ready to execute the contract, all deal points should be spelled out in writing. The information gathered in the RFP response, at the interview, and in the follow-up negotiation should be compiled into a single document forming the basis for the deal. Each deal point should be specific and concise. Attention to detail will significantly reduce the potential for disagreements at a future date.

Critical deal points include the following.

Exceptions/Agreement on Construction Contract

As part of the final deal, it is important to resolve any disputed requirements included in the proposed construction contract. Often an agreement is reached with the contractor early in the design process, and the negotiation of the final contract is left for a later date. In such an instance, the contractor's leverage to change requirements or negotiate a better deal will increase. It is suggested that the following deal point be documented: "Contractor agrees to the contract terms included in the RFP with the following revisions: *(enter or refer to specific revisions if any)*."

Specify Fee Agreement

Details regarding the amount of the fee and the calculation method are of primary importance. Clarify whether it is a percentage or lump sum, whether overhead is included in the fee or is an add-on, and how the fee is calculated in terms of percentage on cost, and specify the fee on change orders.

Define All Markups

All markups to be applied against the total should be clearly identified. Include definition of the markup as well as the amount (percent or lump sum) and how it is calculated. This is often an area for disagreements and is not clearly defined in the AIA Construction Contract, since it is a variable in each project. Markups can include insurance, taxes, safety, bond, preconstruction, and contractor's contingency, as well as others. In the event there is a contractor's contingency included in the deal, be sure to define the method allowing its use during construction (e.g., is owner approval required?) and how any contingency savings may be returned to the owner (or shared) at the time of final accounting.

Agreement on Other Financial Elements

Other financial elements should be addressed. These may include

➤ Bonding requirements

➤ Accounting and audit requirements and procedures

➤ Shared savings agreement and procedures for calculating

> ➤ Specific inclusions and exclusions (in the various markups) such as permits, taxes, home office accounting, reimbursables

> ➤ Performance bonuses

Agreement on General Requirements

To avoid an area of frequent misunderstanding, it is essential that the amount allotted for general requirements (percent or lump sum) be specified, and agreement on what is included as well as excluded be clearly spelled out. The general requirements spreadsheet included in Figure 4-2, coordinated with the CSI MasterFormat™ (Division 01), provide an excellent format to be used for this documentation.

Preconstruction Cost and Extent of Participation

Define the amount (percent or lump sum) and the specific services to be provided. The services may be identified in the RFP and in that case may be specified here by reference. Equally important is to indicate how payment is to be made (e.g., monthly, deferred until construction starts, milestone or performance payment). This is also a good place to define termination and settlement for services in the event the project does not get built or the contractor is not used for construction. Carefully worded terms should be considered for preconstruction termination language, using care not to contradict the contract requirements. Consultation with legal counsel is advised.

List of Committed Personnel

It is often said, "It is the people who make the difference." The importance of documenting the assignment of personnel and roles should not be overlooked. For example, a company may employ only one or two top-notch superintendents, and the remaining employee pool may not include options for personnel experienced in the product type or with a depth of references. The selection process includes meeting the proposed staff and observing the fit with your team, all variables if the proposed staff is changed at a later date.

You may decide to formalize the deal summary and begin work on the project prior to execution of the GMP contract *if* you are confident that the important elements of agreement are in place. Include the RFP and the contractor's proposal in the final deal summary (at least by reference); binding these requirements and commitments into the deal summary may well prove useful at a later date. Use care, though, to clarify specific elements that have been modified or revised during the negotiation. The RFP and proposal often contain requirements and contractor commitments not listed in the contract, or that have been altered or revised in the deal summary. Consider adding a provision that the contract governs in the event of discrepancies.

Another option is to execute the construction contract early, on a "zero-up" basis. A *zero-up* construction contract provides a complete and *executed* contract with price to be entered by change order or amendment. Include language allowing termination and defining any associated penalties or payments. You will then have the means to terminate the contract without undefined penalty in the event that either the project does not go forward or you choose not to use the contractor—obviously preferable to facing mediation or litigation for undefined costs.

In all cases, completion and execution of the formal contract should remain a priority. Construction cost typically accounts for two-thirds of the total project cost. Preparing a clearly documented agreement, defining the elements of the cost of construction as well as the committed resources and definition thereof, is essential to maintaining the construction cost objective and significantly mitigates the potential for scope and cost disagreements. This step is the culmination of the entire RFP process and therefore should be executed carefully and promptly, and with the appropriate duty of care.

CHAPTER 9

What's Next?

Wh_en the deal has been formalized and accepted, you are ready to add the contractor to the team. All the critical elements of the negotiated project delivery team are in place, and the process is set to take off. The complete development team can be brought together to begin the process of project delivery. Architectural design, engineering calculations, pricing, scheduling, and functional decision making can be accomplished efficiently and with confidence of a successful outcome.

The management task now becomes a process of "managing the design to meet the budget." While this may be a lengthy process depending on the size of the project and the related extent of design documents, it is useful to initiate the process with a team meeting, organized to address the project status and establish a road map for success acceptable to all members of the team.

The following topics should be included in the kickoff meeting.

Goal Setting

There are few more important elements of the negotiated-delivery method. Each team member—architect, engineer, consultant, and contractor—has individual and corporate goals. The owner has budget and operational goals, including cost, quality, program, and hold-sell strategies. Goal setting is a process of acknowledging the individual and project objectives, accepting the diversity in these, and finally, and most importantly, melding them into specific project-related goals, providing a framework within which all may succeed.

The negotiated-delivery process is most successful when the owner's goals are presented as binding and guiding principles within which individual goals are met. Buy-in by all team members is essential; and once buy-in is achieved, the goals should be published and distributed. It is best to revisit the goals regularly as the design, permitting, financing, and construction efforts move forward.

Project Definition

The initial kickoff effort should include a detailed project review, to be attended by all key team members, including the civil and geotechnical consultant, architect, structural engineer, mechanical and electrical consulting engineers, interior designer, and other relevant consultants specific to the project type (e.g., acoustic, flooring, water intrusion). Defining the project may be an arduous task, depending on the project size. You may want to assign time slots for certain participants since attendance for the entire session may not be necessary. The Pricing Package from the RFP is an excellent format around which to organize this effort. Plans, reports, and relevant project data should all be reviewed at this initial project definition meeting.

Collaborative discussion of the information is essential to ensure that all team members

➤ Understand the basis of design

➤ Reach consensus regarding design and operational objectives

- ➤ Can voice concerns in a productive environment

- ➤ Become accustomed to the style of other members

- ➤ Have an opportunity to add expertise

Procedural and Communication Parameters

What are the lines of communication? To whom are reports sent? There are so many questions and, if the process is not organized, so much wasted time and effort! The essential organizational chart and brief narrative of responsibilities (if necessary) should be formulated by the team in an interactive process. This will provide an opportunity for the participants to understand the interrelationships and to buy into the plan. The plan should address

- ➤ Organizational hierarchy, which recognizes contractual obligations

- ➤ Roles and responsibilities of key team members

- ➤ Flow of information, down and up

- ➤ Documentation requirements

- ➤ Forms (paper and electronic) required to maintain consistency, provide formal documentation, and simplify distribution

Project Schedule

A review of the critical milestones along the timeline is essential. The discussion will likely reveal key information not previously considered, thereby enhancing the team's understanding of the critical performance issues required to add value to the timeline. Each team member will have tasks to complete to ensure that the milestones will be met. Through collaborative review, participants are informed as to others' roles and tasks and how their own tasks interrelate. This is a tremendous opportunity for paradigm shifting and team building. When the team finally agrees to a milestone schedule, the road map will be complete, guiding the effort to maximum potential for success. Each member will be more likely to take responsibility for meeting her or his respective commitments.

Regular Team Meetings

The best opportunity to organize regular meetings occurs when all active participants are together for the first time. Weekly meetings may be required to assess design progress, value engineer, analyze design and operational decisions, update the budget, and address specific needs and responsibilities of participants, especially those requiring input from others.

Task-specific meetings may also be necessary, requiring limited attendance by those directly involved. These may include finishes selection, mechanical/electrical system decisions, and permit submission strategy. Although additional task-specific meetings may be organized as the design progresses, those that are related to milestones are best scheduled at the kickoff session. The meeting schedule and locations should be published, and every effort should be made to maintain the plan in order to provide participants with an opportunity for predictable and efficient preparation and use of time.

Construction Cost Review

The initial kickoff meeting is a great time to review the elements of the contractor's estimated cost of construction. A detailed review conducted by the contractor provides the opportunity for each team member to

- ➤ Understand the cost constraints

- ➤ Provide feedback regarding the contractor's assumptions

- ➤ Correct inaccurate design assumptions

- ➤ Address cost-saving suggestions

➤ Understand the contractor's estimating format and underlying methodology

➤ Define the construction cost constraints for design maturation

It is an ideal occasion for team building, as it is the first time the contractor leads the discussion. As such, participants should make the effort to recognize the contractor's personnel as essential team members and work to get them up to speed efficiently. The contractor represents a key force in the overall process and is in a unique position to effect a successful outcome. It is wise to establish a respectful two-way working relationship at the outset.

With the contractor at the table, the design process now has a resource for cost, schedule, and constructability input. Individual project components may be studied, analyzed, and finalized within the overall context of meeting the project goals. Design decisions and revisions may be made on a timely basis to complement the design timeline, allowing efficient maturation with reliable cost assurance. The design may be "managed to meet the budget."

Obviously the kickoff meeting may be lengthy and tedious. You may choose to break up the sessions into half-day meetings over successive days or ensure adequate breaks. Allow sufficient time before the meeting for all participants to prepare, and confirm directly with each the props, forms, and other relevant items to be brought to the session. When each team member is adequately prepared, the initial team meeting can lead to an efficient design and construction process with a positive outcome for all. It may very well be one of the most important opportunities to ensure the success of your project.

Summary

This book is intended to provide a broad format for preparing and executing an RFP plan, culminating with contractor selection and a tight deal. Each RFP process will vary according to the team's needs, the project requirements, the stage of document completion, and the intended role of the contractor. The example spreadsheet and forms used in this book must be tailored to fit the RFP plan and the project. As the process unfolds, the plan will require modification to fit reality. Incorporating flexibility into the timing and execution of the plan, as well as the spreadsheets and checklists, will result in an outcome tailored to the specifics of the project.

The RFP effort need not be cumbersome or time-consuming. A well-organized plan, utilizing refined analysis tools, should allow you to complete the entire effort in 3 to 6 weeks. The key steps of a successful RFP process include these:

- ➤ Plan the RFP process in advance.

- ➤ Consider the cost/benefit of the effort.

- ➤ Provide clear and concise information in the RFP document.

- ➤ Complete a detailed analysis of the responses, and follow up as appropriate.

- ➤ Perform thorough reference checks.

- ➤ Conduct open and interactive interviews.

- ➤ Document the negotiation in writing.

- ➤ Execute the deal without delay.

The desired result is the selection of a qualified contractor to assume a critical role in the development of your project—a contractor with the requisite specialized skills to complement your team. Once the selection is complete, essential activities such as value-added decision making, cost/benefit analyses, value engineering, and design-progress budgeting can proceed, using the contractor's current and local market cost database. The proactive interaction of the contractor with the design development team will enhance the probability of your project being built—on time, in budget, and with quality.

Sample Request for Proposal

REQUEST FOR PROPOSAL

Multifamily Project

200 APARTMENTS AND GROUND-LEVEL RETAIL

(date)

For additional information,
please contact:

(insert information)

XYZ CORPORATION, INC.

Table of Contents

SECTION A: Overview

INTRODUCTION

XYZ Corporation is a developer of large multifamily projects in Portland, Oregon. XYZ Corporation plans to develop an apartment project within the city limits of Portland in accordance with this Request for Proposal and Pricing Package.

Your firm has been selected to receive this RFP on the basis of local area research into qualified firms and similar project experience. Five general contractors have been invited to respond.

(insert names)

Response to this RFP is required by *(insert date and time)*, delivered to *(insert vitals)*. The team anticipates a 2-week review period to analyze the proposals. We appreciate your interest in this project.

PROJECT DESCRIPTION

The multifamily project consists of three levels of below-grade parking, small retail spaces at grade, and two wood-framed apartment buildings above a concrete transition deck. The project is located at xxxx Main Street, Portland, Oregon.

The approximate square feet and unit-mix summary are as follows:

➤ Parking (250 stalls):   100,000 gross sq. ft.

➤ Retail at grade: 20,000 gross sq. ft.

➤ Exterior grade-level site development: 45,000 gross sq. ft.

➤ Residential (200 apartments):   220,000 gross sq. ft.

     185,000 net sq. ft.

OWNER'S VENDORS AND CONSULTANTS

The following firms have been selected to join the team:

➤ Architect: _____

- ➤ Structural Engineer: _____

- ➤ Mechanical/Plumbing: _____

- ➤ Electrical Engineer: _____

- ➤ Civil Engineer: _____

- ➤ Landscape Architect: _____

- ➤ Roofing Consultant: _____

- ➤ Construction Manager: _____

Please do not contact these firms for information about the project. Submit all questions through *(insert name of person managing the RFP).*

It is the owner's intention to assemble a team that reflects proactive communication, interactive problem solving, and goal-oriented thinking. You are encouraged to demonstrate throughout the response to the RFP an ability to perform with such a team for the benefit of the owner.

The owner intends to contract with vendors for the following:

- ➤ Telecommunications and data prewire: _____

- ➤ Furniture, fixtures, and equipment: _____

- ➤ Computers and entertainment equipment: _____

MILESTONE SCHEDULE

The key elements of the milestone schedule are as follows:

Date

- ➤ Begin preconstruction and complete schematic design: _____

- ➤ Complete design development and update budget estimate: _____

- ➤ 90% CDs and budget update: _____

- ➤ 100% CDs and GMP: _____

- ➤ Begin construction: _____

➤ Begin retail TI: _____

➤ Beneficial occupancy: _____

The contractor will publish preconstruction and construction schedules identifying key decision milestones, budget updates, permit submission dates, long-lead items, and other critical activities required for the success of the project. The schedules will be reviewed by all team members and revised until such time as team consensus is achieved. The contractor will maintain the schedules and keep all team members informed as to pending decision dates and long-lead order deadlines.

DELIVERY METHOD

The owner intends to execute a Guaranteed Maximum Price contract with the selected contractor, using AIA Document A111™—1997 Version: *Standard Form of Agreement between Owner and Contractor—Cost of the Work Plus a Fee with a Negotiated Guaranteed Maximum Price.* A copy of the intended contract is included with this RFP in Section C.

PROJECT MANAGEMENT/CONTRACTOR REQUIREMENTS

The owner intends to select a contractor who utilizes state-of-the-art project management technology and tools. Such tools should include linked RFI logs, shop drawing/submittal logs, change-order logs, and related project management information databases. Such data should be available to the team members on a project website, accessible by password only. It is desired that communications on the project be electronic in format.

It is intended that the contractor be an essential team member during the entire preconstruction phase. Preconstruction management will include attendance at project design meetings, value engineering, and pricing analysis of alternative systems. The contractor will identify the impact to the construction schedule in all decision matrices. The owner will rely on the contractor to provide accurate and reliable pricing and budget updates to afford the team the opportunity to develop complete documents that meet the budget requirements.

The preconstruction and construction activities will be conducted in an open-book manner. The contractor must be willing to share all pertinent information required to provide the owner with the basis for informed, value-oriented decisions as well as a full understanding of the cost and schedule elements of the project.

The contractor must be willing to review all bids with the owner as well as quantity surveys and material takeoffs. Changes and change proposals on the project must be reviewed with the owner in a similar manner. The contractor's detailed cost reports must be made available to the owner on a periodic basis, no less often than once per month. Substantiated costs backup will be required with each pay application.

The owner is anticipating a base budget will be established with the submission of the response to the RFP. Each budget update will include a comparison to the previous budget so that the team can identify cost changes related to specific elements of the project. Such budget updates will be kept in a UniFormat™ and include detailed supporting backup.

Detailed budget updates will be required at the following milestones:

➤ Schematic design

➤ Design development

➤ 50% construction documents

➤ 100% construction documents: Guaranteed Maximum Price

In the event the GMP amount does not support the pro forma, the owner reserves the right to pursue alternative delivery methods.

During construction, the contractor will hold weekly meetings at the site and will publish minutes of the meetings. Project management reports and logs, the construction schedule, and pending changes will be discussed at each meeting in addition to all relevant project issues. The contractor will monitor all time-sensitive decision requirements to ensure that the owner is informed of the impact of decision deadlines.

The project manager should have a minimum of 10 years of experience in the construction industry with recent experience in similar product types. The project manager will be expected to manage the project throughout preconstruction and construction activities and to be the lead contact for team members. The superintendent should be made available during preconstruction on an as-needed basis and as determined by the project manager and team. Together, the project manager and superintendent should provide the owner with experienced construction expertise to support value-oriented design and construction decisions.

The contractor will commit a principal of the company to participate throughout the preconstruction and construction process. The principal will be available to participate in key project decisions and discussions and will take an active role, following up with the owner to ensure satisfaction with the contractor's performance.

SELECTION CRITERIA

The following criteria will be used in selecting the contractor:

➤ Relevant local experience in similar projects

➤ Qualifications of project manager and superintendent

➤ Competitive proposed fee and markups

➤ Competitive labor rates and burden rates

➤ Demonstrated ability as team player

In the event the responses do not meet the owner's requirements, alternative delivery methods may be pursued.

INTERVIEW SCHEDULE

An interview with your firm may be conducted at the offices of *(insert name)* on *(insert date)*. We will contact you shortly after receipt of the response to schedule such an interview.

SECTION B: Response Format and Definitions

1. RESPONSE FORMAT

The response must be organized according to the following format. Information should be concise and specific to address each request. Include a table of contents and tabs to organize the material in the following order.

2. KEY PROJECT PERSONNEL

➤ Provide an organization chart for the key project personnel, identifying specific individuals. Include resumes for each individual, and list experience with similar product types, if applicable. Provide three references for each individual.

➤ Indicate whether the key team members have worked together on previous projects.

3. KEY FINANCIAL ELEMENTS

➤ Proposed fee for the GMP contract and description of how the fee is calculated and applied against the cost of construction.

➤ Labor rates for self-performed work. Identify separately the associated labor burden rate to be used under this agreement. Include rates for management staff as well as all trade personnel.

➤ Markup for contractor insurance.

➤ Markup for applicable contractor taxes.

➤ Markup for contractor's contingency, if the contractor intends to carry one, stated as a lump sum or percent.

➤ Markup on change orders. (Specify for the contractor and subcontractors, assuming each is different.)

➤ Specify all other markups.

➤ Lump-sum amount, if any, for preconstruction services. It is intended that this amount be incorporated into the final GMP and be payable with the first construction billing cycle. In the event the final GMP does not meet the pro forma and the owner chooses to bid or use an alternative delivery method, or the project is not built, the preconstruction amount will become due and payable to the contractor.

➤ Identify the current backlog for your company and any projects in excess of $5 million currently bidding or negotiating. Include in this section a history of total construction volume for each of the past 4 years.

➤ Submit your proposed plan for shared savings under the execution of a GMP contract. Creative strategies are encouraged.

4. BONDING, INSURANCE, AND FINANCIALS

➤ Provide the overall bonding capacity for the company. Include a letter from the surety indicating the ability of the contractor to bond this project. Identify the bonding rates, and provide history of any claims against bond for the past 5 years. Indicate the duration of the relationship between the contractor and surety.

➤ Provide the contractor's insurance limits proposed for the project. Include a certificate of insurance from the insurance carrier indicating the appropriate coverage. Certificate shall indicate the rating of the carrier and carrier's contact information. Indicate the duration of the relationship between the contractor and insurer.

➤ Provide the most recent certified and audited financial statement for the firm.

5. CONTRACTOR'S COMMITMENT

Provide a letter of commitment on your firm's stationery, signed by a principal. The letter should include (a) the commitment to provide the full faith and credit of the firm in the execution of the work, (b) the commitment of resources and personnel as required to ensure a successful completion of the project, and (c) a commitment to cooperate and coordinate with the owner's agents and contractors at the site.

6. PROJECT MANAGEMENT TOOLS

Provide samples of each of the critical project management tools intended for use on the project. These tools should include, at a minimum,

➤ RFI log

➤ Construction meeting minutes

➤ Submittal and shop drawing log

➤ Sample weekly schedule

➤ Cost reports and updates

➤ Change proposal pricing format

➤ Schedule of values format to be used in pay applications

➤ Overview of quality control program

➤ Detailed safety record for past 2 years

➤ Overview of safety program

➤ Other relevant project management tools

7. REFERENCES

Provide references for the firm. Include three references in each of the following categories:

➤ Architect

➤ Structural engineer

➤ Owner

➤ Building operator or manager

References should represent recent successes in similar product types and be current within the past 4 years. References should support the contractor's proactive team-oriented approach to design and construction.

Provide phone and fax numbers.

8. PROOF OF EXPERTISE

Submit pictures of similar projects and include the following data for each:

- ➤ Name and location

- ➤ Owner (include contact information)

- ➤ Final contract amount

- ➤ Duration and completion date

- ➤ Gross square footage of components (office, retail, residential, garage, other) and associated cost of each component

- ➤ Type of structure (wood, steel, concrete)

- ➤ Awards

9. SELF-PERFORMED WORK

Identify all scopes of work that you intend to self-perform. Include in writing your willingness to bid the self-performed work against at least two other qualified bidders. It is the owner's intention that such scopes of work be bid directly to the owner, and selection be made jointly by the owner and contractor.

10. SCHEDULE

Provide a preconstruction and construction schedule, identifying key elements of the project and key milestones as identified in this RFP. The schedule should be in a CPM format and utilize the same software as intended for use throughout the duration of the project. Identify critical long-lead items and related purchase deadlines.

11. CLAIMS/DISPUTES/LITIGATION

Identify all unresolved and ongoing claims and disputes against your firm in excess of $100,000. Include any claims against the principals of your firm or any claims your company may have against a third party. Provide a history of litigation, including the outcomes, for the past 5 years.

12. COST ESTIMATE

Using the Pricing Package and the documents identified in Section D, prepare and submit (in the UniFormat™) a complete cost estimate for the project. Submit an estimate summary using the spreadsheet provided in Section D. Provide detailed estimate backup for each category. Include a list of qualifications, assumptions, and exclusions.

Contact the construction manager for the format in electronic form.

13. PROJECT COSTS DEFINED

Costs to be included in the estimate are defined in the construction contract. Include a detailed breakdown of general requirements, using the spreadsheet provided in Section D. The contractor is encouraged to expand the spreadsheet to include general requirements not listed, but required to perform the work. The GRs are intended to be complete as required to construct the work. It is the owner's intention to fix the general requirements at the time of award.

Costs for items not specifically defined in the documents should be either qualified or identified as an allowance.

Costs excluded from the estimate should be included in the list of qualifications, assumptions, and exclusions.

It is the owner's intention that all costs required to construct the project ready for beneficial occupancy be included in the estimate.

The following costs will be paid by the owner and are to be excluded from the estimate:

➤ Land use permit

➤ Sales tax

➤ Building permit

➤ FF&E

➤ Paging system

➤ Garage signage

The Pricing Package in Section D provides further definition of project costs.

SECTION C: Construction Contract

A copy of the intended contract is included with this RFP. Your response must enumerate all items for discussion and negotiation that differ from the contract language. It is the owner's intent that the contract be executed within 1 month of award.

Thank you for your effort in preparing a complete response. Please submit five copies for distribution.

We look forward to receiving your response.

Construction Contract

(insert here)

SECTION D: Pricing Package

DOCUMENTS

- ➤ Architectural Drawings
- ➤ Structural Narrative and Drawings
- ➤ Geotechnical Report
- ➤ Environmental Contamination Survey
- ➤ Room Finish Schedule
- ➤ Light Fixture Schedule
- ➤ Plumbing Fixture Schedule
- ➤ MEP Outline Specifications
- ➤ Site Survey

PROJECT NARRATIVES

- ➤ Civil Narrative
- ➤ Architectural Narrative
- ➤ Structural Narrative
- ➤ Mechanical Narrative
- ➤ Electrical Narrative

FORMS

- ➤ Construction Estimate Format
- ➤ General Requirements Format

Project Narratives

OVERVIEW

These Project Narratives are provided to clarify and define the estimate to be submitted by the contractor. Any referenced quantities are not necessarily accurate for use in pricing and should be confirmed by the contractor during the pricing exercise.

Contractors shall carefully examine the documentation for this project to be fully aware of and informed about all the conditions and limitations. If there is any doubt as to the true meaning of any parts of the documentation, submit questions. Submitted questions will be compiled and answered in a timely manner to all contractors. No questions will be taken 3 days prior to the due date of the RFP.

DOCUMENTS

Documents to be used for the pricing effort are enumerated on the previous page. These documents should be used in conjunction with the narratives to assemble an estimated cost of construction organized as indicated in the UniFormat™ provided.

The following narratives provide additional guidance for your pricing effort. It is the responsibility of the contractor to ensure the estimate includes the cost of, and compliance with, these items.

CIVIL NARRATIVE

A. Earthwork

See geotechnical report for recommendations on excavation and site preparation, groundwater management, shoring, structural fill, foundation support, and seismic considerations.

B. Erosion Sedimentation Control

All erosion control shall be per City of Portland Building Department and Bureau of Environmental Services Standards. Erosion control shall consist of the following:

1. Erosion control technical guidance handbook.

2. Construction entrances and truck washdown areas.

3. Bio Bag and/or silt sack inlet filters.

4. Woven filter fabric erosion control fencing.

5. Settling ponds, if required.

6. Additional measures as needed for wet-weather construction.

7. Erosion control blankets on all exposed bio/filtration swales.

C. Water Distribution

1. All public water mains per City of Portland Water Bureau Standards.

2. Pipe materials:

   a. Public: City of Portland Water Bureau Standards (Ductile Iron Cl. 52)

   b. Private: C-900 PVC, Ductile Iron (Cl. 52) Type K Copper, Schedule 80 PVC

3. Water meter size: Estimated two 2-inch domestic services per City of Portland Water Bureau Standards in vault with sump pump. Installation of an approved double check valve assembly in vault with sump pump will be required directly behind the water meter.

4. Fire service size: Estimated two 8-inch services. Double check detector assembly will be required in vault with sump pump, and with fire department connection.

D. Storm and Sanitary Pipe

1. All stormwater will be discharged to the existing storm system in the public right-of-way after treatment and detention.

2. Treatment: Project will likely use a stormwater management vault system or downstream defender–type manholes.

3. Detention: Detention will be required, and a location for a detention facility will be a challenge. An estimated volume of 5000 cubic feet of storage will be required for the two blocks. This can be handled by using utility vault three-sided bridge systems (two 20-foot × 25-foot × 5-foot areas), but finding a location for this structure within the site is questionable. Other options could include possible rooftop detention or a piped detention facility in the pubic right-of-way, if allowable by the City of Portland. The contractor shall allow $100,000 for detention.

4. Storm pipe size (10-inch service for each building):

   a. Hancor Hi-Q HDPE

   b. ADS N-12

   c. SDR Rated HDPE

   d. ASTM C900

5. All public sanitary sewers installed per City of Portland Building and Bureau of Environmental Services Standards. Public lines in easements and in public right-of-way.

   a. Sanitary sewer pipe (10-inch service for each building): D3034 PVC

E. Sanitary and Storm Miscellaneous Structures

1. Vault detention system: Utility vault company three-sided bridge system or equal. System shall meet requirements of BES exhibits 3-9, 3-9b, and 3-10.

2. Water quality system: Stormwater management vaults or approved equal.

F. Subdrainage

1. Footing drains will be included for the buildings with appropriate backwater valve connection to site storm system.

2. Under-slab drainage has not been identified as a requirement by the geo-technical report other than standard groundwater management during construction (i.e., dewatering, etc.). No cost should be included.

G. Flexible Pavement

1. Street paving within the existing adjacent public right-of-way is required within 10 feet of property line.

2. All asphalt paving shall be per City of Portland Transportation Department Standards.

3. Aggregate base material shall conform to City of Portland Transportation Department Standards.

H. Pavement Marking

1. Section includes: traffic lane and parking stall striping, handicap symbol painting, and directional-arrow painting: AASHTO M-248, Type 3F

I. Cast-in-Place Concrete Curbs

1. Assume that all curbing along all street frontages will be replaced.

2. Portland cement concrete shall be 3000 psi in 28 days.

3. Preformed expansion joint: conform to AASHTO M-153 or AASHTO M-213.

4. Curing compound: White pigmented curing compound shall conform to the requirements of ASTM C-309.

J. Concrete Sidewalks and Driveways

1. All concrete sidewalks and driveway aprons within the public right-of-way will be constructed to details and specifications of the City of Portland. Contractor is responsible for coordination of inspection by the city's Public Works Department for this work. It should be assumed that all existing street sidewalks will be removed and replaced as a part of this project.

K. Irrigation System

1. The contractor shall assume that irrigation will be required for all planting components of the project including street trees and roof terracing. Drip systems with point source emitters will be the design intent rather than open spray heads. A separate meter will not be provided for the irrigation system. A 2-inch stub will be provided after the domestic meter, and then a backflow device will be required for the irrigation system. The landscape contractor will be responsible for installing all hardware after the meter. The general contractor will be required to coordinate installation of the domestic meter based on the civil engineering plans.

2. All irrigation equipment and installation methods shall conform to the standards established by the City of Portland.

3. Obtain and pay for any permits and inspections required by governing agencies and utility companies.

4. Provide a complete working system with full head-to-head coverage.

5. Coordinate power supply requirements with the electrician.

6. All equipment shall be Rainbird or approved equal.

7. All lateral pipes will be class 200 PVC pipe.

8. All pressurized pipe including mainline will be Schedule 40.

9. All underground sleeving will be Schedule 40.

10. As-built will be required at completion, indicating dimensional data.

L. Landscaping

1. All underground parking for this project will be to the back of the street curb and also underneath 13th Avenue. Due to this fact street trees will need to be installed in concrete vaults, and bracing will need to be provided to support the vault. Bracing calculations will be provided by a structural engineer. A drainage system will be required in the bottom of each vault to drain water from the vault. The exact method of disposing of this storm drainage is yet to be determined; however, removal of excess water from these pits will be required. A rooftop plaza will also be installed for entrance to the residential developments. Planting will occur on this second level primarily in pots and raised planters, to provide spatial arrangement and separation between the units. The contractor should anticipate the placement of large deciduous trees (minimum 2-inch caliper), small specimen trees, and an irrigation system on this upper level.

2. All planted areas shall conform to the standards established by the City of Portland. Street trees will be 3-inch caliper.

3. A soils test will be required to determine the suitability of the soil for backfill mix. Soil test should indicate amendments to be placed in the soil.

4. Tree wells will be excavated to a depth 24 inches below finished grade. The plant pit will be filled with $2/3$ existing topsoil to $1/3$ amended topsoil. Of the $1/3$ volume, 30% will be humus. The humus and $2/3$ and $1/3$ topsoil will all be mixed together and placed back into the pit.

5. Provide trees, shrubs, and other plants complying with the recommendations and requirement of the American Nurserymen's Association.

6. Provide stakes for all trees made of cedar or redwood, free of knotholes and other defects. Provide wire ties, guys, and black two-ply garden hose to protect tree trunks from damage by wires.

7. All landscaped beds will be covered with a 2-inch (minimum) depth of medium dark fir bark mulch if in an area suitable for mulch, and not pea gravel as mentioned below.

8. All street trees will be planted in a 4-foot × 6-foot tree well with tree frame and tree grate. A 2-inch layer of pea gravel will be placed in tree well as a cover mulch.

End of Civil Narrative

ARCHITECTURAL NARRATIVE

A. Coordination and Inclusions

1. Contractors are to familiarize themselves with the requirements of all governing ordinances of codes or other authorities having jurisdiction relating to hours and protection of public, use of streets, protection against fire, and any other conditions which may be required.

2. Design-build stairs and all related components.

3. Coordinate slab, courtyard, and roof opening with all disciplines.

4. Coordinate and sequence the preparation and processing of shop drawing submittals with the performance of the work so that the work will not be delayed.

5. Provide all temporary construction facilities with all required temporary utilities for the completion of work.

6. Maintain at the site all project record documents.

B. Demolition

1. Include the complete removal and disposal of all existing buildings, foundations, floor slabs, and all related items. Remove debris found on site and resulting from operations as it accumulates.

C. Exterior

1. Precast parapet cap at retail, and precast sills at all windows.

2. Brick veneer: 4 × 4 × 16 brick veneer with 2-inch airspace, building paper over $5/8$-inch exterior glass fiber–faced gypsum sheathing over structural metal wall framing; provide self-adhering membrane flashing at all building paper transitions and around all openings.

3. Storefront: $1 3/4$-inch × 6-inch clear anodized front offset system manufactured by Kawneer, US Aluminum, or Arcadia. Provide 14-foot 0-inch × 8-foot 0-inch automatic sliding door packages for block one retail and 3-foot 0-inch × 8-foot 0-inch medium stile single and double doors as shown for block two.

4. Exterior metal panel system: 22-gage, G90 galvanized steel with polyiso-cyanurate urethane foam core, smooth finish with kynar 500/hylar 5000.

5. Deck railing systems: System 1: $1/2$-inch tempered plate glass panel with metal rail clip system. Manufacturers: Hansen Architectural Systems, Tubular Specialties Manufacturing, Inc. (TSM), Livers Bronze Co. System 2: Cable railing system. Manufacturers: Hayn Enterprises LLC, Johnson Architectural Hardware, Seco South Tension Systems.

6. Raised paving system: Provide raised paving and pedestal system with drain matting and all required components at the plaza level. Waterproof membrane to be continuous under planters, stem walls, stairs, and other structures. The plaza levels are defined as the level above the retail areas, the first residential floors.

7. Skylights: Provide custom metal and clear glass skylight canopy. Manufacturer: Bristolite, Tristar, or Acralight. Provide a wash and drain system for cleaning.

8. Exterior metal canopies: Provide $1 1/2$-inch metal decking over miscellaneous metal framing with 10-inch metal channel surround, hold back decking 3 inches for front drainage, provide 1-inch-diameter rod supports to building.

9. Roofs: Built-up roofing system of four-ply, asphalt-applied, membrane system with granule surface finish. Include all flashings and sheet metal work, flashing of all roof penetrations and roof drains.

D. Conveying Systems

1. One elevator: 4500-pound, speed 150 fpm freight car with 4-foot 0-inch doors retail.

2. Four elevators: 2100-pound, speed 350 fpm with 3-foot 0-inch doors for residential.

E. Interiors

1. General

   a. All interior nonbearing partitions to be $3 5/8$-inch metal studs @ 16-inch OC with $5/8$-inch gypsum board each side. Provide type "X" at rated walls. Provide sound attenuation insulation at bathroom and bedroom walls.

b. Provide furred down ceilings to 8-feet 0-inches at all rooms except living and dining rooms, bedrooms, and dens. Use $5/8$-inch gypsum board over metal joist framing.

2. Interior finishes

   a. Carpet

      i. Apartment units (all locations except where other flooring is specified): Carpet to be 22-ounce minimum face weight, 12-foot-wide rolled goods, cut pile, carpet pad 4-pound density, rebind pad $3/8$-inch thickness. Approved manufacturers: Shaw, Lee, J & J Commercial.

      ii. Public spaces (corridors and elevator lobby): Carpet to be 22-ounce minimum face weight, 12-foot-wide rolled goods, tufted loop, direct glue-down. Approved manufacturers: Shaw, Lee, J & J Commercial.

   b. Resilient flooring

      i. Apartment units (kitchen, bathroom, W/D closet): Resilient flooring to be vinyl composition tile, 12 inches × 12 inches. Approved manufacturers: Azrock, Armstrong, Mannington.

      ii. Public spaces (all ancillary spaces, i.e., common bathroom, janitor closets, mechanical room, storage room, etc.): Resilient flooring to be vinyl composition tile, 12 inches × 12 inches. Approved manufacturers: Azrock, Armstrong, Mannington.

   c. Cabinetry

      i. Apartment units: Cabinets to be frames: $3/4$-inch plastic laminate over substrate; doors: plastic laminate. Approved manufacturers: Lanz Cabinets, Canac, Diamond Cabinets.

      ii. Public areas (common kitchen, bathroom): Same as apartment units.

   d. Countertops

      i. Apartment units (bathroom, kitchen): Countertops to be laminate, with 4-inch backsplash. Approved manufacturers: Wilsonart, Nevamar.

      ii. Public areas (common kitchen, bathroom): same as apartment units.

e. Paint—walls

   i. Apartment units (all rooms): Paint to be light texture, no or low VOCs, two coats over primer required. Approved manufacturers: Benjamin Moore, Sherwin Williams, Miller Paint Co.

   ii. Public areas (all rooms): Paint to be light texture, no or low VOCs, two coats over primer required. Approved manufacturers: Benjamin Moore, Sherwin Williams, Miller Paint Co.

f. Base

   i. Apartment units (all rooms): Base to be rubber straight base, 4 inches high. Approved manufacturers: Roppe, Flexco.

   ii. Public areas (all rooms): Same as apartment units.

g. Doors

   i. Apartment units (all rooms): Doors to be wood hollow-core, painted. Approved manufacturers: Sun-Dor-Co, Lag Design, Industrial Millwork Corporation.

   ii. Public areas (all rooms): Doors to be wood, painted, flush, solid door, 20-minute rating. Approved manufacturers: Sun-Dor-Co, Lag Design, Industrial Millwork Corporation.

h. Lighting

   i. Apartment units: Bathroom sconces, dining room pendant, closets downlight. Approved manufacturers: Progress, Eureka.

   ii. Public areas: Corridors, sconces, or indirect downlighting on walls. Approved manufacturers: Progress, Eureka.

i. Hardware

   i. Apartment units (all rooms): Hardware to be chrome finish. Approved manufacturers: Schlage, Corbin.

   ii. Public areas (all rooms): Hardware to be chrome finish. Approved manufacturers: Schlage, Corbin.

j. Staircases

   i. Apartment units: N/A.

   ii. Public areas: Staircases to be rubber treads. Approved manufacturers: Roppe, Flexco.

k. Wallcovering

   i. Apartment units: N/A

   ii. Public areas: N/A

l. Closets

   i. Apartment units: Shelving to be white melamine—one hat shelf per closet.

m. Bathroom shower unit

   i. Apartment units: Shower unit to be fiberglass enclosure. Approved manufacturers: Best Bath, Fiberfab.

End of Architectural Narrative

STRUCTURAL NARRATIVE

A. Substructure

1. The foundation system will consist of concrete spread footings. Footing sizes 8 and 12 feet square, per plan.

2. The perimeter below-grade basement walls will be shotcrete. Wall thickness is 10 inches. The lowest level will be a 4-inch-thick concrete slab on grade.

3. Temporary shoring will be required for construction of the below-grade parking levels. Include cost per square foot identified in the RFP.

B. Superstructure

The superstructure will consist of several different systems depending on the level of the building. The following summarizes the proposed structural system for each level:

1. Below-grade parking levels

    a. The floor framing system will consist of post-tensioned concrete flat slabs spanning between cast-in-place concrete columns. The slabs will be 7.5 inches thick.

    b. Column grid spacing will be approximately 30 feet. The columns will be rectangular with a width of 16 inches and depth of 20 inches. Stud rails are required.

    c. The lateral load-resisting system will consist of concrete diaphragms spanning between concrete shear walls. The shear walls will align with retail and residential-level concrete shear walls above and will be 14 inches thick and located around the stair/elevator cores and at selected interior locations. The perimeter basement walls will also resist lateral loads.

    d. At several locations in the parking garage, concrete transfer beams will be required to address offsets in column locations. Allow $50,000.

2. Street level

   a. The floor framing system at the retail and parking areas will consist of post-tensioned concrete flat slabs spanning between cast-in-place concrete columns. The slabs will be 8.5 inches thick. Column grid spacing will be 30 feet.

   b. Columns at the retail area will be 18 inches square.

   c. Columns at the parking area will be rectangular with a width of 16 inches and a depth of 20 inches.

   d. The lateral load-resisting system will be the same as that for the below-grade parking levels.

3. Podium level (residential first floor)

   a. The floor framing will consist of a post-tensioned concrete flat slab spanning between cast-in-place concrete columns. This will be a transfer slab 14 inches thick.

   b. The lateral load-resisting system will consist of a concrete diaphragm spanning between concrete shear walls. The shear walls will be 12 inches thick and be located around the stair/elevator cores and at selected interior or perimeter locations. Walls will align with shear walls above.

4. Residential levels

   a. The floor and roof framing will consist of 8-inch cold-formed metal joist at 16-inch OC spanning between 6-inch metal stud bearing walls. A 1/2-inch-deep metal form deck will span between the metal joists and support a 3-inch-thick normal-weight concrete slab.

   b. The lateral load-resisting system will consist of concrete floor and roof diaphragms spanning between shear walls. The shear walls will be cast-in-place concrete. The shear walls will likely be 10 inches thick and be located around the stair/elevator cores and at selected interior and perimeter locations.

C. Material Specifications

1. Allowable soil bearing

   a. Dense sand 4000 to 5000 psf

   b. Dense gravel 6000 to 10,000 psf

2. Concrete

   a. Slabs on grade: 3000 psi

   b. Foundations and basement walls: 4000 psi

   c. Columns and shear walls: 5000 psi

   d. Parking and podium PT slabs: 5000 psi

   e. Concrete slabs on metal deck: 4000 psi

3. Reinforcing steel

   a. ASTM A615, Grade 60, deformed bars.

4. Post-tensioned reinforcement

   a. $1/2$-inch-diameter, 270-ksi unbonded monostrand, fully encapsulated.

5. Cold-formed steel framing

   a. Studs and track:

      i. 16-gage and heavier shall be ASTM A446 Grade D, or ASTM A570 Grade 50.

      ii. 18-gage and lighter shall be ASTM A446 Grade A, or ASTM A570 Grade 33.

D. Design Loads

1. Wind: 80 mph exposure "B"

2. Seismic: Zone 3, soil type Sd

3. Roof: 25 psf snow plus drift

4. Floors:

    a. Exit corridors   100 psf

    b. Retail   100 psf

    c. Retail mezzanine   125 psf

    d. Residential   40 psf

    e. Parking   50 psf

    f. NW 13th Avenue   HS-25

    g. Sidewalks   250 psf

End of Structural Narrative

MECHANICAL NARRATIVE

A. HVAC Systems

 1. Parking garage HVAC

 a. Parking garage exhaust will be sized for an approximate effective exhaust rate of 0.75 cfm/sf using the alternate method of ventilation provided by the UBC. Exhaust will be taken at the walls of the garage on both levels. A single-plug fan will serve the block 1 side of the garage and will be located in a fan room on the P1 level. It will exhaust to a wall louver above grade. The block 2 garage will be served by single fan located on the P1 level with discharge to a wall louver above grade.

 b. The exhaust fans shall be provided with variable-frequency drives. Carbon monoxide sensors shall modulate the exhaust fan speed for energy-efficient operation. Makeup air shall be drawn down the entry ramp (provide a grated, not solid, door).

 c. Provide exhaust and electric unit heaters for the fire pump room and domestic water service room. Provide exhaust for the electric service room and main telephone/data room.

 d. Fans shall be belt-driven with housing, wheel, fan shaft, bearings, motor, and mounting brackets. Cook, Greenheck, or approved.

 2. Retail space HVAC

 a. Retail space will be served by a water source heat pump system consisting of heat pump units, cooling tower, boiler, pumps, and hydronic piping loop.

 b. A forced-draft cooling tower with a nominal capacity of 150 tons shall reject heat from the hydronic water piping loop. The cooling tower will be located at the P1 level. Discharge from the cooling tower will be ducted to a wall louver above grade. Cooling tower shall be blow-through and counterflow. Baltimore Air Coil, Evapco, Marley, or approved.

 c. One natural gas–fired boiler with an input of 1000 MBH shall provide heat to the hydronic water distribution loop during the heating season. One hydronic water pump will be provided for circulation. Boilers shall have stainless steel burners and be factory-assembled and tested. Raypak, Tele-dyne Laars, or approved.

   d. Air handling equipment and water source heat pumps shall be factory-assembled and manufactured by Trane, Carrier, York, or approved.

3. Apartment HVAC

   a. Apartments will not be provided with mechanical cooling. Electric resistance baseboard heaters will provide heat. Ventilation air will be provided via operable windows.

4. Apartment bathroom exhaust

   a. Bathroom exhaust will be sized at 6 air changes per hour. High-quality, low-noise ceiling exhaust fans, connected to the light switch or separately switched (by Division 16), will be provided for each of the bathrooms. The exhaust fans will be ducted to an exhaust shaft riser consisting of a 2-hour rated shaft with a sheet metal exhaust duct riser (do not use shaft as exhaust duct). Each exhaust duct connection will be provided with a 22-inch subduct to eliminate the requirement for a fire damper or fire smoke damper at the shaft penetration. The exhaust riser and roof-mounted exhaust fans will be sized assuming a 75% diversity (i.e., 75% of the exhaust fans are on at one time). The roof-mounted exhaust fans will be driven by variable-frequency drives to adjust fan capacity to maintain a constant shaft negative static pressure. Emergency power is necessary to meet the code for continuous operation.

   b. Fans shall be belt-driven with housing, wheel, fan shaft, bearings, motor, and roof curb. Cook, Greenheck, or approved.

5. Apartment kitchen exhaust

   a. The kitchen exhaust hood (provided as part of the appliance package) will be a recirculation type with carbon filters.

6. Apartment dryer vents

   a. Dryer vents will be connected to an exhaust riser with a 22-inch subduct similar to the bathroom exhaust risers. Sizing is based on 100 cfm per dryer. The exhaust riser and roof-mounted exhaust fans will be sized assuming a

50% diversity (i.e., only one-half of the dryers are on at one time). Each riser fan will be equipped with a variable-frequency drive to adjust fan capacity to maintain a constant shaft negative static pressure. Emergency power is necessary for code-required continuous operation. A cleanout at the lowest level of each dryer riser is to be provided for duct cleaning.

    b. Fans shall be belt-driven with wheel, housing, fan shaft, bearings, motor, and roof curb. Cook, Greenheck, or approved.

7. Corridor HVAC system

    a. Corridor conditioning will be provided from a 1800 cfm packaged air handler on the roof. Gas-fired heating and DX cooling are provided for conditioning. A mixed-air economizer section allows operation with 100% fresh air for energy savings and for slight pressurization.

    b. Units shall be factory-assembled and tested. Trane, Carrier, York, or approved.

8. Retail tenant HVAC

    a. Future retail tenants will be served from water source heat pump units provided with water from the hydronic distribution system. Fresh air for ventilation will be provided from outside air intake louvers installed in the exterior face of the retail spaces. Exclude the cost of heat pumps (by tenant).

B. Plumbing

1. Waste and vent

    a. Waste and vent systems: cast iron waste risers with separate vent lines.

    b. Oversize washing machine drains to mitigate sudsing.

    c. A 2000-gallon grease interceptor will be provided for the retail areas.

    d. A 10-inch sewer main is anticipated.

2. Storm drainage

    a. Storm drainage will be routed in cast iron no-hub pipe from the drains on the upper roof. Some of the drainage will be routed to lower roof planting areas to reduce the amount going to the municipal system. Drains in the planting area will be provided to take care of any excess water. This runoff will be routed through a storm interceptor provided under the civil work. Two 10-inch storm drain mains are anticipated.

b. A separate piping system will be used for the overflow drains. These will be routed down to discharge 18 inches above grade.

3. Domestic water

   a. Type L copper with soldered or brazed joints for domestic hot and cold water piping systems.

   b. The main domestic cold water risers may be victaulic joints to accommodate shop fabrication and field installation with mechanical couplings.

   c. Distribution in units will have a local shutoff valve and PEX tubing manifold. A 6-inch main is anticipated.

4. Natural gas

   a. Natural gas will be provided for the boilers, makeup air handlers, and central domestic water heater systems.

   b. Gas service shall be centrally metered with separate meters for retail.

   c. Gas piping shall be Schedule 40 steel pipe, with welded fittings. Branch piping may be screwed for connection to ranges.

   d. A 4-inch main is anticipated.

5. Piping insulation

   a. Fiberglass insulation shall be provided for domestic hot and cold water piping systems.

   b. Horizontal rainwater leaders shall be insulated with fiberglass blanket.

6. Pressure zones

   a. Provide a triplex (5 hp each) booster pump system with variable-speed drives to maintain a consistent water pressure for the upper units.

b. Shutoff valves will be provided in each unit and in the top floor corridor to allow shutdown of individual risers for servicing.

c. Manufacturers: Paco, Grundfos.

7. Domestic water heating systems

a. Provide a central domestic water heating system using the two 750 MBH gas-fired boilers on the roof with 600-gallon storage tanks and piping distributed throughout the building.

b. The domestic hot water piping would parallel the domestic cold water piping including the upper and lower stations and pressure zones. Alternate: Provide a 50-gallon, 6-kW electric water heater for each unit.

c. The lower floors will be supplied hot water from four condensing-type water heaters (equivalent to Cyclone BTH 250) located in the booster pump room on level P1.

d. Manufacturers: A.O. Smith, Weben Jarco, Ray Pak.

8. Plumbing fixtures

a. Residential quality plumbing fixtures are anticipated throughout each unit. Final selections will be by architect/owner's marketing team.

b. Manufacturers: Kohler, American Standard, Eljer.

C. Fire Protection

1. System shall be design-built by a licensed contractor. Design documents shall include flow and pressure calculations as required to support issuance of permit. Permit shall be included in the cost. Designer shall utilize Auto-CAD and coordinate design with others.

2. General
The entire facility including the penthouse, all units and corridors, the retail spaces, and parking structure is to be protected by an automatic fire protection system in accordance with NFPA 13. A combination standpipe and fire protection riser is proposed for one of the exit stairwells with a zone valve serving each level. The other stairwell will be provided with a wet standpipe

and fire department hose connections. Pipe routing through the units shall be carefully coordinated with the structure, especially where exposed slabs are provided and sprinkler lines must run in soffits or through metal stud walls serving sidewall sprinkler heads. The parking garage system shall be dry, and piping shall be galvanized for corrosion control. The main header should be looped around the core below the beam line, with branch piping running up between the beam pockets to maximize floor height.

3. Fire pumps

    a. The height of the building will require a fire pump system to pressurize the risers, to meet the pressure requirements of the upper floors of the tower.

    b. The fire pump room will be located at the parking level and will include a 25-hp electric fire pump and 3-hp electric jockey pump to maintain pressure at all times in the system.

    c. Manufacturers: Peerless, Fairbanks Morse, ITTAC.

End of Mechanical Narrative

ELECTRICAL NARRATIVE

A. Power Distribution Systems

1. Utility distribution and equipment

   a. Existing street lighting, traffic control signal, and trolley catenary poles will be affected by the project excavation. Street lighting upgrades required by the city will be paid by the owner.

   b. Existing utility poles along Marshall Street have pole-mounted transformers and serve properties across Marshall Street to the north of the site.

   c. Existing transformer and switch vaults are installed in the sidewalk along 12th Avenue.

   d Services to other properties will be relocated by the utility, paid by owner.

   e. New service transformers will be installed in transformer rooms at level P1 and will be paid for by the owner.

   f. One transformer will step the utility primary voltage down to 480Y/277V for service to the retail tenants, while a second transformer will step the utility primary voltage down to 208Y/120V for the residential tenants.

   g. Utility metering provisions will be included at the main switchboard for the retail tenants and at residential meter stacks for the residential tenants.

   h. Scope of work includes design and construction of the service laterals from the vaults to the building's service equipment, as well as coordination with the utility to schedule and sequence utility work around the site, including relocation of utilities for other affected properties.

   i. Design and construction of utility vaults and primary raceways will be performed by the utility.

2. Main services

   a. The project will have three services as follows:

      i. One service for retail shell spaces.

ii. One service for parking structure loads.

iii. One service for residential tenant loads.

b. The 2000-amp retail service switchboard lineup shall be sized to accommodate three large (400- to 1000-amp) 208Y/120V, three-phase services and three or more smaller (less than 200-amp) 208Y/120V, three-phase services.

c. The residential service will be 3000-amp, 208Y/120V, three-phase switchboard lineups sized to accommodate all residential loads. These switchboards will then subfeed residential meter stacks located in electrical rooms on upper residential floors.

d. The building shall be metered with utility company meters for retail service, house service, and residential service. The retail services shall be separately metered within the switchboards. The retail and house service switchboards will be provided with utility incoming section and metering sections as required by the utility. The residential services will be separately metered at the residential meter stacks.

e. Manufacturers: Square D, Cutler-Hammer, General Electric.

3. Normal power distribution

a. Retail distribution: From the retail service switchboard, feeder conduit only shall be routed to the retail tenant space for use during the tenant improvement phase of construction. The voltage and ampacity of these services and feeders shall be determined by the owner and accommodated in the design. Allow $50,000.

b. House distribution: "House" panelboards will be provided in the main electrical room within the parking garage, in electrical closets on the residential floors, and in the mechanical penthouse for convenience outlet, lighting, and miscellaneous power requirements. House loads will include, but not be limited to, the following: retail lobby heat pumps, retail lobby AC units, HVAC control panels, fire/smoke dampers, electric water coolers, building supply fans, building return fans, toilet and dryer exhaust fans, and elevator equipment.

c. Parking dstribution: Panelboards will be provided in the main electrical room within the parking garage for fans and pumps and will subfeed a lighting and appliance panelboard to accommodate convenience outlets and lighting. Parking loads will include but not be limited to, the following: sump pumps, sewage ejectors, garage exhaust fans, and electric heat trace as required for freeze protection.

d. Residential tower distribution: 208Y/120V, three-phase electrical risers will run from the services in the main electrical room to 208Y/120V, three-phase meter stacks on the residential floors. Two risers will serve units on the 1st through the 3d residential floors, while two other risers will serve the 4th and 5th residential floors. Individual 208/120V, one-phase feeders will connect the residential meter stacks to tenant panelboards within each residential unit. The panelboards within each unit will feed branch circuits to lighting, appliance, and receptacle outlet loads within the unit.

e. Panelboards shall be manufactured by the same manufacturer as the service switchboards.

4. Alternate power systems

   a. Emergency generator

      i. A 300-kW, 480Y/277V, three-phase diesel engine-generator to provide auxiliary power for emergency (NEC 700) and legally required standby (NEC 701) loads will be located in a generator room.

      ii. A skid-mounted day tank shall provide fuel for the emergency generator, and an exhaust muffler and piping system will route exhaust outside.

      iii. The requirement for a rated room will be confirmed with the code Authority Having Jurisdiction (AHJ) during schematic design.

      iv. Manufactured by Caterpillar, Onan, or Kohler.

   b. Emergency riser

      i. The generator set will serve an emergency riser for emergency loads such as egress lighting, fire pumps, emergency generator loads, and other loads as defined by the AHJ.

      ii. Emergency panelboards will be distributed vertically in the electrical closets in each building to accommodate loads.

c. Standby riser

    i. The generator will serve a standby riser for legally required loads such as toilet and dryer exhaust fans and other loads as defined by AHJ.

    ii. Standby panelboards will be distributed vertically in the electrical closets in the high-rise along with normal and emergency house panels.

5. Convenience power outlets

    a. Power outlets shall be provided to meet all applicable codes and standards, located as follows:

        i. In all common areas such as lobbies, hallways, and restrooms.

        ii. In all utility spaces such as mechanical, electrical, and elevator equipment rooms.

        iii. Throughout penthouse mechanical rooms to allow proper servicing of all equipment.

        iv. At all telecommunications backboards.

    b. The maximum number of convenience outlets on one circuit will be six. Maximum loading of lighting branch circuits will be 60% (i.e., 12 amps). Dedicated circuits will be provided for telephone board and specialty areas.

    c. Within the residential units, convenience power will be provided with receptacle outlets in units as required by and spaced per the NEC. Each kitchen, family room, dining room, living room, bedroom, or similar room or area of the dwelling units shall have the minimum quantity of receptacle outlets as required in NEC 210 plus 1.

6. Miscellaneous requirements

    a. In addition to Division 15 mechanical equipment, elevator equipment, convenience outlets, and lighting, power connection will be provided for miscellaneous equipment including, but not limited to

        i. Roll-up door at loading dock

        ii. Window washing equipment at roof

iii. Parking attendant booths

iv. Handicapped access doors at entries to building

vi. Trash compactor at loading dock

vii. Building manager's office and PC workstation and telecommunications systems

B. Lighting Systems

1. General

   a. High-efficiency lighting will be provided for the building. Because the project exceeds four stories, the design shall comply with the requirements of the Oregon Non-Residential Energy Code.

2. Within other interior spaces lighting will be provided as follows:

   a. Stairwells and corridors: wall-mounted fluorescent, 20 footcandles (fc).

   b. Mechanical rooms, electrical rooms: strip fluorescent, 20 fc.

   c. Retail shell spaces: strip fluorescent "stumble lighting," 5 fc. Finished lighting will be part of the retail tenant improvements (TI).

   d. Parking garage lighting will be provided as follows:

      i.   Lobbies: downlights, cove lights, and decorative sconces, 20 fc.

      ii. Stairwells: wall-mounted fluorescent, 20 fc.

      iii. Parking: surface and wall fluorescent strip lights mounted to meet the following luminance criteria:

Area	Day[a]	Night[a]	Uniformity[b]
Interior parking and drive areas	6	6	3:1
Minimum at perimeters	2	2	
Vehicle entrances[c]	50	6	3:1
Vehicle exits[c]	25	6	3:1
Ramps and corners	15	15	3:1
Stairwells and pedestrian exits	20	20	3:1

[a]Indicates minimum maintained horizontal luminance in footcandles at 30 inches above finish floor.

[b]Indicates average to minimum ratio.

[c]Entrance and exit areas defined as from edge of building 50 feet into the building.

3. Site lighting

    a. Egress and architectural lighting shall be provided on the exterior of the building within canopy areas, entrance courtyards, and doorways. Up/down fluorescent sconces will be spaced approximately 20 feet on center. Each unit's courtyard shall have one illuminated bollard. This work shall be coordinated with the architect. Roadway and public sidewalks shall be illuminated under a separate contract.

    b. Egress lighting: Egress lighting will be a subset of the general building corridor lighting system. Egress lighting will be designed to comply with the City of Portland, Bureau of Buildings, Program Guide, "A Guide to Procedures and Requirements, Egress Lighting," dated June 1, 1996. Exit signs will be high-efficiency LED type.

4. Lighting controls

    a. All enclosed spaces will be provided with local wall switches for lighting control.

    b. All exterior lighting shall be routed through an exterior lighting control panel, consisting of contactors, a time clock with reserve power, and a photocell on the roof. Individual contactors shall be provided for exterior lighting zones 1 through 4, outlined below.

    c. At exterior entrance doors, a minimum maintained light level of 2 fc for a distance of 20 feet from entrances will be provided.

    d. Exterior lighting shall be controlled by a "dusk-to-dawn" (photocell on–photocell off) circuit.

    e. Light source shall be low-glare, metal halide luminaries.

f. At the building facade, the light source shall be low-wattage metal halide. Facade lighting shall be controlled by a "dusk-to-preset off" (photocell on–time clock off) circuit.

C. Telecommunication Systems

1. Voice communication

   a. The main telecom room shall be located at level P1. The design shall include all infrastructure including the main telecom room; telecom closets on ground, 1st residential, and 5th residential floors; necessary risers from the main telecom room to telecom closets; pathways to all tenant and house spaces; and wiring to outlets as indicated above. Redundant conduit feeds will be provided from the main telecommunications room for use by the multiple service providers.

   b. Conduits from the main telecom room will be provided through the parking structure wall and will be stubbed out and capped at the property line. All conduit stubouts shall have watertight seal and metallic location tape installed. The telecommunication entrance services and pathways will be coordinated with the designated service providers.

   c. Additional conduits will be routed from the main telecom room to the telecom closets located on the residential tenant floors, with junction boxes required to accommodate continuous wiring from the ground level to uppermost telecom closet.

   d. Raceways will be provided as required by code for emergency communications within the elevators, and a raceway system will be provided for outlets for house phones in the lobby.

   e. A conduit will be provided from the telecom closet located on the residential tenant floor to each retail tenant's space. Under tenant improvement work, data and telecommunication closets for each of the retail tenants will be provided within the tenant space.

   f. A raceway system will be provided for outlets in all occupied spaces within the units with the exception of bathrooms. The locations will be determined by the architect.

2. Cable TV distribution

a. A cable TV system riser conduit will be provided as part of the building conduit riser system for future cable TV services. The cable TV entrance services will be coordinated with the designated service providers.

b. Conduits from the main telecom room to the telecom closets located on the residential tenant floors, with junction boxes required to accommodate continuous wiring from the ground level to uppermost telecom closet, will be provided.

c. A conduit will be provided from the telecom closet located on the residential tenant floor to each retail tenant's space.

d. Conduits between the telecom closets located on the residential tenant floors and the roof for future satellite and microwave dishes will be provided.

3. Broadband data communication

a. High-speed Internet raceways and outlets will be provided in all occupied spaces within the units with the exception of bathrooms. The locations will be determined by the architect.

b. System will include Category 5E (Cat 5E) wiring with dual-jack RJ45 outlets. Within the units, outlets will be provided in all occupied spaces with the exception of bathrooms. The locations will be determined by the architect. This raceway system will also provide outlets at the building manager's office for its PC workstation.

D. Life Safety System

1. A state-of-the-art, microprocessor-based life safety system will be provided. It will include smoke detection, water flow detection, alarm annunciation, stairway and exit door unlocking, and magnetic door holder release. The system shall comply with City of Portland Fire Department and Bureau of Building's requirements, ADA, and UBC life safety system requirements, including "area of rescue assistance" system.

E. Security Systems

1. A raceway system for CCTV cameras at lobby, exterior courtyard, and garage elevator stop will be provided. Security system equipment will be provided and installed by others.

2. Similarly, a raceway system for an access control system will be provided for an entrance card key and call button (scrolling display and full apartment intercom/door release) systems. Access control system shall be provided and installed by others.

End of Electrical Narrative

Construction Estimate Format

Note: Contact RFP manager for electronic format.

Total Project Gross SF	340,000	SF
Net Rentable SF	185,000	SF
Number of Units	200	EA

UniFormat BREAKDOWN	Total	% of Subtotal	$/GSF
STRUCTURE		0.00%	$0.00
EXTERIOR CLOSURE		0.00%	$0.00
ROOFING		0.00%	$0.00
INTERIOR CONSTRUCTION		0.00%	$0.00
CONVEYING SYSTEMS		0.00%	$0.00
PLUMBING SYSTEMS		0.00%	$0.00
HVAC SYSTEMS		0.00%	$0.00
FIRE PROTECTION		0.00%	$0.00
ELECTRICAL SYSTEMS		0.00%	$0.00
SITE WORK/DEMOLITION		0.00%	$0.00
GENERAL REQUIREMENTS		0.00%	$0.00
SUBTOTAL		0.00%	$0.00
Contractor's Fee		0.00%	$0.00
State & Local Taxes		0.00%	$0.00
Insurance		0.00%	$0.00
Contractor's Contingency		0.00%	$0.00
Other		0.00%	$0.00
TOTAL CONSTRUCTION COSTS			
Escalation		0.00%	$0.00
Bond		0.00%	$0.00
Preconstruction Amount		0.00%	$0.00
Other		0.00%	$0.00
TOTAL			$0.00

General Requirements Format

Note: Contact RFP manager for electronic format.
Add lines to include additional items.

XYZ CORPORATION BUILDING Description	CONTRACTOR A Total SF: 340,000	
	$	$/SF
PROJECT STAFF & SUPPORT		
Project Manager	—	0.00
Project Superintendent	—	0.00
Project Engineer	—	0.00
Cost/Scheduling Engineer	—	0.00
Office Administrator	—	0.00
Travel & Subsistence	—	0.00
Project Legal Fees	—	0.00
Professional Engineering Services	—	0.00
Scheduling	—	0.00
Total Project Staff & Support	—	0.00
PROJECT SAFETY/SECURITY		
Safety Training	—	0.00
Safety Engineer	—	0.00
Security/Watchman	—	0.00
Security Equipment	—	0.00
Total Project Safety/Security	—	0.00
FIELD SUPERVISION & FIELD ENGINEERING		
Field Supervision	—	0.00
Field Engineering	—	0.00
Engineering Equipment	—	0.00
Engineering Supplies	—	0.00
As-built Documents	—	0.00
Quality Control	—	0.00
Certified Survey	—	0.00
Total Field Supervision & Field Engineering	—	0.00
PROJECT OFFICE		
Office Equipment	—	0.00
Office Supplies	—	0.00
Telephone/Data	—	0.00
Postage & Shipping	—	0.00
Copies/Reproduction	—	0.00
Drinking Water	—	0.00
Project Office Setup/Dismantle	—	0.00
Project Office Rental/Maintenance	—	0.00
Copy Machine	—	0.00

(CONTINUED)

XYZ CORPORATION BUILDING	CONTRACTOR A Total SF: 340,000	
Description	**$**	**$/SF**
PROJECT OFFICE (cont.)		
Computers	—	0.00
Printers & Fax Machine	—	0.00
Furniture & Equipment	—	0.00
Total Project Office	—	0.00
TEMPORARY FACILITIES & SUPPORT		
Trailer Rental	—	0.00
Temporary Fence & Maintenance	—	0.00
Shop Costs	—	0.00
Tool & Dry Sheds	—	0.00
Project Signs	—	0.00
Temporary Guardrails	—	0.00
Safety Equipment & Supplies	—	0.00
Weather Protection	—	0.00
Temporary Stair Towers	—	0.00
Temporary Ladders	—	0.00
Temporary Fire Protection	—	0.00
Protect Finishes	—	0.00
Temporary Roads/Parking	—	0.00
Street Sweeping/Dust Control	—	0.00
Street and Use Permits	—	0.00
Traffic Control/Barricades	—	0.00
Dewatering	—	0.00
Test Equipment	—	0.00
Temporary Office Janitorial	—	0.00
Total Temporary Facilities & Support	—	0.00
TEMPORARY UTILITIES		
Install Temporary Power Service	—	0.00
Temporary Power Bills	—	0.00
Temporary Power Distribution	—	0.00
Temporary Lighting	—	0.00
Temporary Water Bills	—	0.00
Temporary Water Distribution	—	0.00
Portable Toilets	—	0.00
Temporary Heat	—	0.00
Total Temporary Utilities	—	0.00
HOISTING/FORKLIFTS		
Mobile Crane Rental	—	0.00
Mobile Crane Operator	—	0.00

(CONTINUED)

XYZ CORPORATION BUILDING	CONTRACTOR A Total SF: 340,000	
Description	**$**	**$/SF**
HOISTING/FORKLIFTS (cont.)		
Forklift Rental	—	0.00
Forklift Operator	—	0.00
Fuel & Maintenance	—	0.00
Total Hoisting/Forklifts	—	0.00
CONSTRUCTION EQUIPMENT		
Construction Equipment (attach detail)	—	0.00
Fuel & Maintenance	—	0.00
Small Tools	—	0.00
Consumables	—	0.00
Shop Foreman	—	0.00
Freight & Trucking	—	0.00
Total Construction Equipment	—	0.00
PERSONNEL/MATERIAL HOISTING		
Protect Elevator Doors & Fronts	—	0.00
Protect Elevator Cabs	—	0.00
Man/Material Hoist	—	0.00
Hoist Operator	—	0.00
Elevator Maintenance	—	0.00
Elevator Operator	—	0.00
Total Personnel/Material Hoisting	—	0.00
CLEANUP		
Continuous Cleanup	—	0.00
Final Cleanup	—	0.00
Disposal Bills	—	0.00
Window Cleaning	—	0.00
Total Cleanup	—	0.00
PROJECT CLOSEOUT		
As-builts	—	0.00
Operation & Maintenance Manuals	—	0.00
Total Project Closeout	—	0.00
Totals:	**0.00**	**0.00**
Notes/Comments:		

Index

About the Author

Richard T. Fria owns, operates, and is president of The Fria Company, Inc., which provides project and construction management services on large commercial developments. He is well-known for on-time and in-budget delivery while assuring adherence to program and aesthetic requirements.

Mr. Fria has worked in the industry for 35 years, completing over five million square feet of projects totaling in excess of $500 million. His expertise in office, biotech, hi-tech, retail, multifamily, and hospitality construction provides a diverse understanding of complex designs, systems, and budgets. His team-oriented leadership and management style have contributed to award-winning success, including

➤ AGC Award for Excellence in High-Rise Construction and Canada Pre-cast Industry Project of the Year for the Second & Seneca Building

➤ AGC Grand Award for Excellence in Construction for Pacific Place

➤ AGC Grand Award for Excellence in Construction, NAIOP Development of the Year, and AIA Renovation of the Year (Honorable Mention) for the ZymoGenetics Building

Drawing on his experience working with nationally recognized design firms, developers, and owners, he has formulated unique techniques for "managing design to meet budget and program," particularly in the negotiated project delivery approach. The RFP is a key element in this process.

In this book, Mr. Fria presents a systematic tutorial on getting the most from the negotiated construction RFP process. The system is designed to deliver extensive, project-specific information to the owner and design team, affording an ideal opportunity to validate schedule and cost assumptions as well as to collect relevant data useful in refining the scope, design, and budget. This process offers a quantitative and thorough means to determine which contractor team will add the most value to the project.

For access to free downloads of forms and templates that can be customized for your specific RFP situations, please visit http://books.mcgraw-hill.com/engineering/updatezone.

The Fria Company (www.friaCM.com) contracts to developers, owners, operators, and architects, and provides services ranging from full-service project management to task-oriented functions such as managing the RFP, strategic project planning, value engineering, quality assurance, construction and project management, and problem-solving. Special emphasis is placed on team play and value-added consensus decision-making.